ROBOT IS THE BOSS

HOW TO DO BUSINESS WITH
ARTIFICIAL INTELLIGENCE

ROBOT IS THE BOSS

HOW TO DO BUSINESS WITH ARTIFICIAL INTELLIGENCE

ARTUR KIULIAN

ISBN-10: 0-692-94540-7
ISBN-13: 978-0-692-94540-7

TABLE OF CONTENTS

INTRODUCTION

"Just as about 100 years ago electrification changed every single major industry, we're in the phase where AI will now do the same. Almost anything that a typical person can do with less than one second of mental thought we can either now or in the very near future automate with AI."

– Andrew Ng, Google Brain founder, and co-founder of education startup Coursera

This book is not about how Artificial Intelligence (AI) will destroy humanity or how machines will rebel against us. Instead, it explains the best way to benefit from using machine learning in your business today. It's not technical; it's simple. But why is it needed? Aren't there already a lot of books on the topic?

If you search Amazon.com for AI books, you won't see a single book that is written for an entrepreneur to take action on the trend of machine learning. Most of the books are either "Surviving AI Revolution" books aiming to capitalize on the fears of SkyNet taking over, or super technical writings like "Large Scale Machine Learning with Python" for developers and data scientists.

There are many books about technological singularity and books from famous futurologists, such as Ray

Kurzweil, trying to reverse engineer the human brain. There are even books tackling the ethics of using AI and robots. But none of those books can tell you how to introduce machine learning to your company.

You would assume there would be press coverage of all these trends including interviews with influencers about the topic. There is indeed much press and buzz about startups with fancy .ai domain names. But there is zero actionable business insight for entrepreneurs across the hundreds of blogs and publications in the AI world.

With so much fluff that sounds like science fiction instead of actionable information, entrepreneurs and business owners are overwhelmed. Who wouldn't be? Too many questions come up unanswered each time you read that a Japanese firm replaced 600 workers with an algorithm.

That is the biggest problem our business society is facing right now. Everyone knows the storm is coming and everyone knows what the storm is about, but only few can get the necessary things done.

Is it too late to start using AI, or is too early? Is this just another buzzword media is cashing in on through Clickbaits? Why is this happening and how can I as an entrepreneur follow the trend? This book is aimed to answer all those questions and guide you through the complex topics with easy-to-grasp language.

PART 1: IDENTIFYING THE PROBLEM

CHAPTER ONE

WHY ARTIFICIAL INTELLIGENCE IS BECOMING SO IMPORTANT

This section details the current machine-learning revolution and how important it is for the future of work and business.

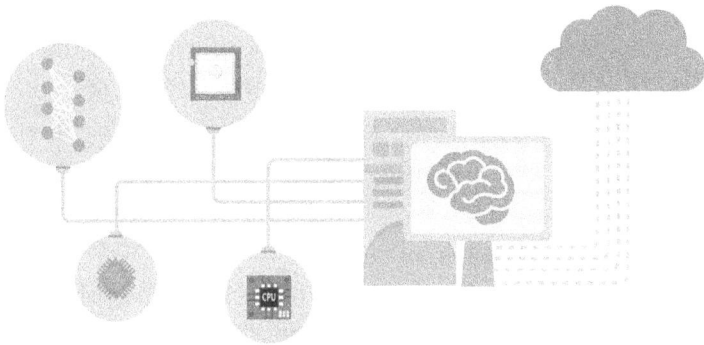

Recent advances in AI, or machine learning, play an increasing role in our everyday life. As a result, its potential effect on the workplace has become a major focus of research and public concern.

Why should you care about the trend? Even though robots and automation across the industrial landscape are not exactly a new thing, it's worth mentioning that they are coming to our everyday work lives soon from burger-flipping robots replacing humans in restaurants, to online chatbots replacing thousands of support call centers. These recent advances enable a new generation of systems that exceed human capabilities across a variety of tasks.

These new systems can learn to do jobs in a matter of minutes that take humans years to master. They are powered by large amounts of data that no human could ever grasp. But they are not superhuman intelligent because those machines are not conscious (at least for now). They operate on a very limited scale of designated tasks.

These machines may soon know more about your family than you do, may know more about your eating habits than you ever notice, and already know when it's time to wake up to avoid traffic like the Google Pixel phone software does.

It's hard to grasp, but AI is already here, in almost every big app or company whose services you use. For example, Google Maps navigation learns from the past experiences of millions of drivers, and Airbnb shows intelligent price recommendations when you list your apartment for rent.

It's easy to feel that you and your workplace are behind technologically since every major company already benefits from AI and are grabbing the low-hanging fruit.

But that's not true; we are only at the beginning of the adoption cycle which is powered by a growing demand.

So why now? What exactly happened that triggered such an extensive expansion into automation and smart machine adoption?

The answer lies in the data availability and growing accessibility of computational power. If you have no idea what this means, that's okay; that's exactly why you are reading this book.

CLOUD PROCESSING

Cloud computing has been one of the most talked about subjects in the tech industry for more than 15 years but is now finally reaching its full potential, which is very important for AI. When talking about cloud computing, it's worth mentioning that the costs of cloud computing are going down as fast the processing power goes up. This advent of cheap online processing and storage is explained by Moore's Law – the rule of thumb that the density of transistors on a microchip doubles about every two years. Fortunately, this rule is still true even after 60 years, which makes prices go down consistently each year.

One of the biggest benefits of cloud computing for organizations is the ability to maintain infrastructure at required levels without technically owning physical servers. This reduces costs by paying for only what is required when it's required.

Another benefit of cloud computing is the ability to scale when needed without increases in management and administration overhead. Cloud computing frees up company IT departments to focus on their strategic objectives instead of worrying about server infrastructure.

For some large companies, AI is becoming not only essential but existential. Joaquin Candela, head of the Facebook Machine Learning Team, said in a recent interview, "Facebook today cannot exist without AI. Every time you use Facebook, Instagram, or Messenger, you may not realize it, but your experiences are being powered by AI."

According to Sundar Pichai, CEO of Google, the company is evolving from a mobile-first to an AI-first world. Therefore, their cloud infrastructure is being optimized around machine learning.

The involvement of big corporations is making progress go even faster. Amazon, Google, Microsoft, and IBM have already rolled out next generation AI computing products in their existing services such as Amazon Web Services, Google Cloud Platform, Microsoft Azure and IBM Bluemix.

This booming cloud infrastructure of corporate giants is becoming widely available for anyone to experiment with machine learning for almost nothing. Things that took researchers years and cost millions of dollars back in 2005 cost almost nothing now.

TALENT

Inexpensive computing and never-before-seen infrastructures create opportunities for anyone willing to take part. A huge pool of talent from university researchers to early stage startups are driving more and more accomplishments in the field.

The number of AI and Machine Learning degrees being obtained worldwide is higher than ever before. More than 170 universities currently offer programs, and more are available online.

Companies such as Udacity and Coursera provide training in machine learning, deep learning, data science, and self-driving specializations that cover all the needed information to get hired by the top companies within the industry.

Udacity, for example, has built hiring partnerships with companies like Mercedes-Benz, Nvidia, BMW, McLaren, Bosch, Amazon, and Intel. Udacity's program is a new funnel for a highly-specialized talent trained through so-called nanodegrees based on a truncated field of study that spans months, rather than years, and allows students to direct the pace of their own learning.

Having these big companies as hiring partners proves the value of Udacity's programs to potential students since they are exactly the types of companies at which students are looking to get jobs.

Most of this self-learning happens online, so students can learn from anywhere without the need to relocate. This fuels another strong aspect powering the talent revolution: globalization.

Companies are no longer limited to hiring scientists and engineers locally. You can efficiently work with people spanning across different time zones and get the benefits of hiring the most appropriate people for the job even if they are not sitting next to you.

The number of events, conferences, and even hackathons, on the subject of AI is rapidly increasing. Google "Top 20 AI and Machine Learning Conferences for Developers This Year" to find an event near you.

DEEP LEARNING

It's not necessary to cover the entire history of how AI has evolved over the last 60+ years, but you should be aware of many recent discoveries.

In the past couple of years, AI has become truly capable of satisfying the needs we force it to, such as speech recognition interfaces, ordering products from Amazon via Alexa, and searching for all the dog photos on your iPhone in one second. Algorithms are now capable of detecting breast cancer. Self-driving cars can pick you up from a restaurant, or park in your driveway.

One of the main breakthroughs to thank for these quantum leaps is a mainstream application of deep learning: a method to discover more intimate relationships between undiscovered in advance aspects of data, to the extent of eliminating human teacher from the equation.

To fully understand deep learning, it's important to digest some terminology which may sound confusing at first.

Artificial intelligence by itself is a broad term that describes various technologies used to mimic or exceed human intelligence. Machine learning is a narrow term that is used to describe specific software techniques that allow computers to learn from the previous experiences and data. Deep learning, scientifically named deep neural networks, is a method of machine learning that is powered by learning to represent the world as a nested hierarchy of concepts, with each concept defined in relation to simpler concepts.

ARTIFICIAL INTELLIGENCE

MACHINE LEARNING

DEEP LEARNING

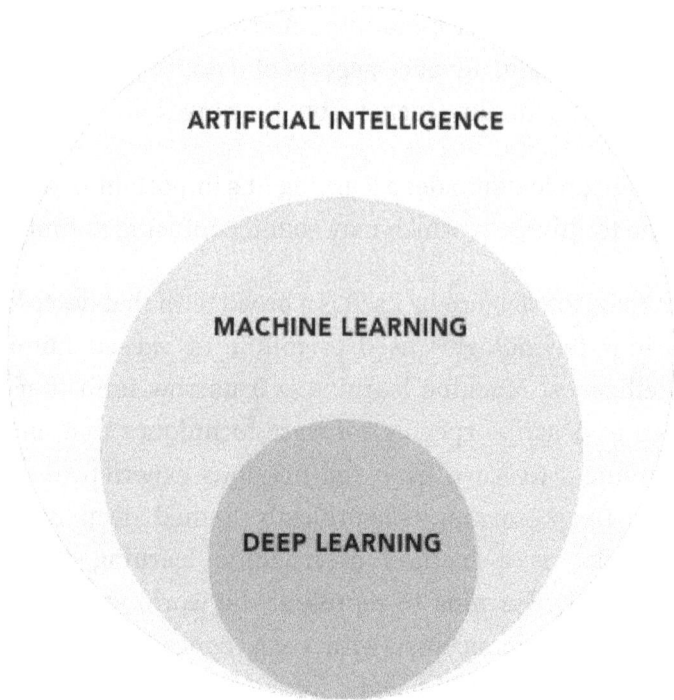

The most remarkable thing about deep neural nets is that no human has programmed a computer to perform any of the predefined actions. And, no human could do that, in contrast to more classic machine learning approaches where a human can handcraft some of the features of the algorithm.

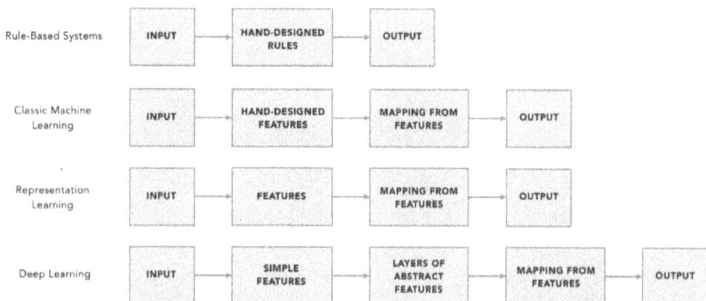

Rule-Based Systems	INPUT →	HAND-DESIGNED RULES →	OUTPUT	
Classic Machine Learning	INPUT →	HAND-DESIGNED FEATURES →	MAPPING FROM FEATURES →	OUTPUT
Representation Learning	INPUT →	FEATURES →	MAPPING FROM FEATURES →	OUTPUT
Deep Learning	INPUT →	SIMPLE FEATURES →	LAYERS OF ABSTRACT FEATURES →	MAPPING FROM FEATURES → OUTPUT

Instead, a deep neural network is composed of multiple layers of neurons that are exposed to terabytes of data to train on. This learning allows algorithms to figure by itself how to recognize traffic signs, how to separate dogs from cats, and how to detect the signs of cancer within the normal cells.

Though the concept of a machine teaching itself and neural networks is nothing new, these algorithmic breakthroughs go back to the 1980s. What's different now is the fact that those algorithms can work at full capacity, with the abundance of computation power and data available across the Internet.

Most current data is represented digitally, which means if a deep neural net is trained on photos, it uses pixels as a raw digital input. Because it is difficult for a computer to understand the meaning of a raw sensory input, the deep learning approach resolves this difficulty by mapping certain pieces of the input into smaller details. So, the first layer of a neural network detecting cats would represent little edges, the ones that represent dark sides with separated bright sides. The next layer of neurons combining the data from the previous layer would learn to detect things like corners where two edges form an angle. One of these angles may represent the shape of the cat's ears, for example.

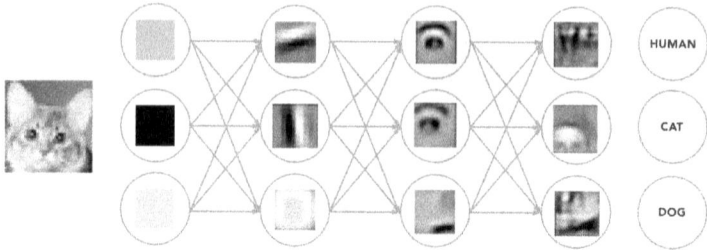

The next level may find more complicated visual structures, such as angles arranged into a circle that may represent an eye, or a specific neuron may represent the cat's nose. Similar to how humans can detect objects simply by seeing pieces of the picture, the algorithms are being trained to compose a full picture from these multi-layered representations of a raw pixel data. Each layer represents concepts of greater complexity and abstraction until one of the top neurons correspond to a full concept from the desired training set, such as a cat.

This is the typical way deep neural nets construct reality from data, but the most important thing here is the need for a neural net to see how right it is in guessing the final representation. That is how testing data sets are introduced for a neural net to send signals down the layers and tweak its parameters to improve the results and learn.

This is how the Google Translate tool accidentally created its own language that is basically a digital representation of all the human languages inside the concepts that correspond to specific words and phrases. This type of abstraction allows efficient translation to any language in a matter of milliseconds.

HARDWARE

Another major recent catalyst for progress in AI is the use of graphics processing units (GPUs) for training large neural network models instead of using traditional CPUs. There is a big structural difference in how GPUs and CPUs process tasks since CPUs consist of a few cores optimized for sequential serial processing and GPUs have an architecture optimized for parallel processing that consists of thousands smaller, more efficient cores. This type of architecture allows GPU cores to handle multiple tasks at the same time.

Chipmakers such as NVIDIA designed GPUs to render images for games and other highly graphical applications for visuals to be rendered faster since those are computed from thousands of pixels changing every single second. This fast and parallel architecture that can handle multiple tasks concurrently is perfectly suited for the type of training computation neural networks are doing since they process enormous amounts of high dimensional data.

By 2011 AI researchers around the world had discovered GPUs which allowed the famous Google Brain project to achieve unseen before results. The project unexpectedly learned to spot cats and people in videos after being trained on millions of YouTube videos. Though this fun experiment required an enormous amount of resources, its processing infrastructure has been handling over

2000 CPUs in one of the Google's giant data centers. Few have had computers of this scale.

Bryan Catanzaro with NVIDIA Research and Andrew Ng's team at Stanford joined forces to use GPUs for deep learning. The results of their research concluded that 12 GPUs could deliver the same deep-learning performance of 2000 CPUs. Researchers at NYU, the University of Toronto, and the Swiss AI Lab have also accelerated their deep neural networks on GPUs.

Since then, GPUs became the shovels to the AI gold rush. Along with GPUs enhancements, there is a big part of industry working on a so-called AI specific chips, chips designed to accelerate AI type of computations.

In May 2016, at their Developers Conference, Google shared that they have been using an internally-developed processor called a Tensor Processing Unit (TPU) to accelerate Deep Learning applications. TPUs already power many Google apps including RankBrain (to improve the relevancy of search results) and Street View (to improve the accuracy of maps and navigation).

TPUs also powered AlphaGo in matches against Go world champion, Lee Sedol, enabling it to think much faster and look farther ahead between moves.

Google has reported that they have run TPUs inside data centers for more than a year. During this time, these machines have delivered a better-optimized performance per watt for machine learning, which equates to fast-forwarding technology seven years into the future.

TPUs are customized to machine learning applications which allows computation to squeeze more operations per second into the silicon, use more sophisticated and powerful models, and apply these models more quickly to give users more intelligent results faster.

Other chip makers such as ARM are racing to create better computing architecture. ARM recently unveiled its next generation of processor microarchitecture named Dynamiq. Chips built using this processor will be easier to configure which allows more powerful systems-on-chip, as well as processors that better serve future computing tasks from artificial intelligence to self-driving cars. According to ARM, Dynamiq will deliver a 50x increase in "AI-related performance" in the next five years.

Intel has made moves towards this trend by acquiring Israeli startup Mobileye, which makes chips and software for self-driving cars. And even Apple is making moves towards optimizing their on-phone chips to provide faster and more optimized computation to the end device. Which is super exciting and means much better user experience that will cause more engagement with the end products which in turn will increase amounts of data generated and holistically improve the algorithms through optimization. We can only imagine the scale of these innovations coming in through the next 5-10 years and how much better hardware will become due to all this chip race competition.

DATA MONSTERS

In addition to having lots of computational power, the key to building intelligent machines is including the right algorithms and proper data. Without those, we can only build simple machines capable of limited tasks, such as playing checkers. And that's exactly why more and more companies are focusing on data harvesting, especially across business processes, to craft better algorithms.

DATA IN THE WORLD vs DATA STORAGE COSTS

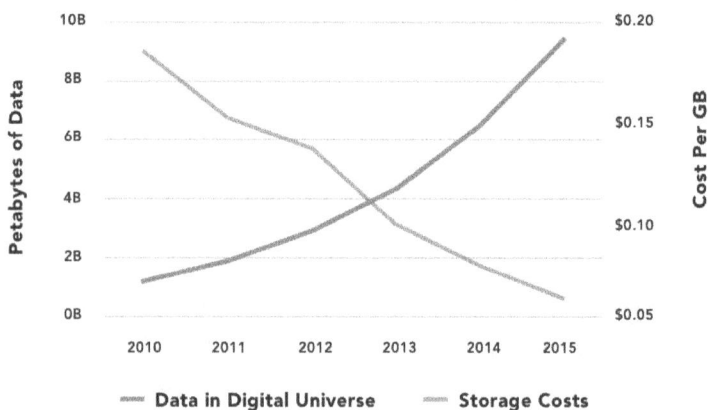

Source: KPCB INTERNET TRENDS 2016

Luckily, large amounts of data are generated every second. For example, retail giant WalMart processes more than million customer transactions every hour which fills databases with more than 2.5 petabytes of new data. This is more data than is stored in the whole America's

Library of Congress. And it's happening everywhere. The total amount of data stored globally is measured in millions of terabytes, and that is doubling as fast as computation power does. There are more than 350,000 tweets posted each minute on Twitter and Facebook by itself is home to more than 50 billion photos.

Data is generated on the Internet through website visits, Google searches, emails, advertising responses, social media posts, likes, form signups, etc. Even more data is generated in the offline world when we carry our GPS-equipped smartphones, wear smartwatches, exercise at the gym, or use our credit cards at stores.

We leave digital footprints in forms of bytes everywhere we appear. Sensors continuously capture real-time operations in factories, hospitals, vehicles, aircraft and other industrial environments.

All these ever-growing piles of data are being viewed as resources that can produce enormous amounts of value, now and in the future. Just as data-intensive companies like Facebook are generating profits from stocking up the data about its user base, any other company can organize the data around their business.

Each of these micro interactions is generating much more valuable data that one can imagine. There are more than 4 million Facebook post likes generated each minute, which means each like creates some kind of representation of what a specific Facebook user looks like, which specific posts this user likes, which specific topics and pages he or she likes, which ads this user responds to, the average click-through rate based on this single interaction, the best demographic of users that responds to this image, which ad performs better during this time of the day, which drives more conversions to the website that is serving ads, and which page leads to a better conversion. An infinite amount of data is generated by a simple click or Like.

Most of this data is "unstructured", which means it is captured in a wide variety of formats that simply don't stack nicely together in one place or database. Humans have adapted to this type of unstructured data coming in from a real world, but we are very bad at scaling that perception when it comes down to terabytes of data tables across different domains. This is where machine learning and so-called big data are playing a big role.

Machine learning powered by data has the potential to transform the nature of knowledge-based jobs within the industry. The predictions can be extracted from the data and will be increasingly used to substitute for the human capabilities such as experience and judgment. The layers of middle management workers will likely disappear, and many of the jobs that are now done by skilled analysts will transform into the ones served by the more powerful data hungry machines.

RUSHING CAPITAL

These dramatic advances and progress across the AI industry have sparked a burst of investment activity.

According to CB insights research firm, equity funding of AI-focused startups reached an all-time high of 658 deals in 2016, up from 160 deals within a four-year period. Over 550 startups raised five billion dollars in 2016.

"Venture capitalists, who didn't even know what deep learning was five years ago, today are wary of startups that don't have it," says Frank Chen partner of Andreessen Horowitz venture fund.

Larger corporate companies have acquired many private companies working to advance AI. These acquisitions and start-up activity ensure AI will continue to be a hot funding item in VC for the next few years. If you are looking for early stage financing, you can easily find 2400+ investors interested in AI just by looking through the AngelList site or any other investor community.

AI ANNUAL GLOBAL FINANCING HISTORY

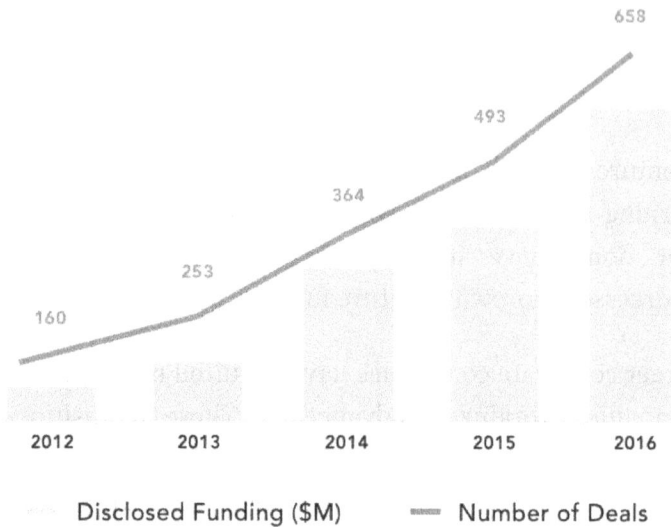

Disclosed Funding ($M)   Number of Deals

Source: CBINSIGHTS

What this means is that there is no shortage of resources and money for the new companies exploring this space. It also means that there is more money being invested to establish the so-called infrastructure products that will make it even easier to build intelligent applications on top of in the future.

JOB MARKETS

It's important to review history to understand the potential impacts and how innovations affected workers and job markets. Even though The Industrial Revolution made society better overall, the transformation led to severe disruptions to the work lives of many people.

Many families were forced to move to new communities to acquire new skills and find new jobs.

Technological changes sometimes substitute for some skills while complementing others. The trends change over time. There were periods of time in history when new technologies raised productivity and increased employment opportunities for workers with little education and other times new technologies increased employment opportunities for high-skilled workers. For example, in the 1800s, looms and machinery replaced the jobs of highly-skilled textile weavers earning high wages with a combination of machines and lower-skilled labor. The high-skilled workers were no longer valuable in the market, but more opportunities were available for less-skilled workers.

In the 20th century, technological changes had an opposite effect on the job market. Computers and the Internet improved the relative productivity of higher-skilled workers. New technologies replaced many routine task-based occupations such as switchboard operators, filing clerks, travel agents, and assembly line workers.

The demand for college-educated labor to complete non-routine cognitive tasks increased as technology advanced. Since college-educated employees were already more highly compensated, the rising demand and increase in their relative pay further contributed to rising pay inequality.

What this means for business is that skills and high levels of education will be the focus for the workforce. And most importantly, the shifting nature of the professional requalification will start affecting the low skilled layers of workers as the global economy evolves in ways that will make it more difficult for people with lower levels of education to find jobs and support themselves. This will eventually force job markets to demand more and more high-skilled labor and education. These changes will affect many businesses and steer competition for high-skilled talent.

THE FUTURE OF WORK

It is not a simple task to predict exactly which jobs will be immediately affected by the oncoming AI automation since machine intelligence is not a single technology. It is a whole group of technologies which applications differ from specific tasks to specific industries, like computer vision technology in security pace can be totally different from manufacturing type of machine vision. Therefore, these effects of oncoming machine intelligence will be felt unevenly through the economy leading to some types of tasks being easily automated and some not which will cause some jobs to be affected more than others.

But what's clear is that effects of AI in the decade ahead will continue the trend toward skill-biased change that computerization and communication innovations have started.

Humans will still maintain a comparative advantage over AI and robotics in several areas. AI can detect patterns and create predictions, but cannot replicate social or general intelligence, creativity, or human judgment. High-skilled occupations using these types of skills will require higher levels of education.

Despite what is being said in recent press, AI revolution is not necessarily about replacing human workforce. It's about people and machines working together to improve the work we do, the new type of partnership our society has to embrace. And the future of such partnership will be divided along the lines of what AI can and can't do.

It will always be a challenge to create AI that can truly replicate the kinds of human abilities we take for granted, things like creativity or human interaction. The complexity of human relationships, minds, and work cultures is currently beyond the scope of what's possible. That's why the proper augmentation of human's work with machine learning and other technologies is the safest bet.

Companies of the future need to differentiate decision making not just regarding its result, but also according to its uniquely human aspects. Some of the tasks workforce does daily is considered creative and unique when it's not. Some things are considered to be easy to automate when in reality things will break right after real human comes out of the interaction. The definitive goal for a business is to properly assess and identify those time-consuming, non-mechanical tasks and produce more time for the workforce to focus on truly valuable pursuits. So, a lot of the art and skill in figuring out where to insert AI is to recognize the business opportunities where you have a complex system that consists of one-second tasks that might be easily tied together automatically.

AI will reduce the time spent looking for information because it will have the information. The future workplace will need to ask the right questions. Executives will shift from a traditional emphasis on procedural knowledge to creative thinking and problem solving, with AI as a research assistant.

Another important aspect will be the notion of feasibility; the fact that automation is possible does not mean it will be practical or cost effective for the business itself. In some cases, human labor will remain less expensive and more effective for at least the foreseeable future.

TECHNICAL FEASIBILITY

% of time spent on activities that can be automated by
adapting currently demonstrated technology

Source: McKinsey Research

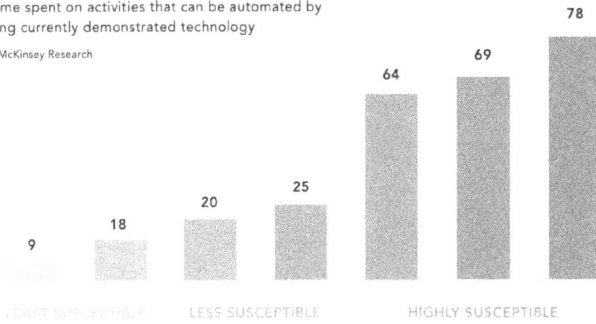

9   18   20   25   64   69   78

LEAST SUSCEPTIBLE   LESS SUSCEPTIBLE   HIGHLY SUSCEPTIBLE

TIME SPENT IN ALL US OCCUPATIONS, %

7   14   16   12   17   16   18

APPLYING EXPERTISE   UNPREDICTABLE PHYSICAL WORK   DATA PROCESSING

MANAGING OTHERS

STAKEHOLDER INTERACTIONS   DATA COLLECTION

PREDICTABLE PHYSICAL WORK

CHAPTER TWO

BENEFITS OF ARTIFICIAL INTELLIGENCE

This section explains the benefits of using Artificial Intelligence (AI) and the long-term effects of neglecting automation trends.

Using Machine Learning and AI in business provides many benefits. Machines achieve high and consistent levels of production without needing breaks or time off like humans.

Machines are scalable and more accurate than humans. In many industries, the removal of human error and the introduction of laser-like precision at faster rates greatly increases profits. Using the right combination of AI and related technologies provides the advantages of reliability and cost-effectiveness while addressing uncertainty and speed.

Maximization of value output and minimization of manual error is the Holy Grail of any business. But it's a tough thing to accomplish using human labor. Any big company approaching the future era of automation without automated processes in places is doomed for extinction. Every little piece that seems to work with putting an extra human on top of it will eventually cause a breakdown on a scale. This lack of efficiency is simply economically unpredictable and urges companies to follow the automated economies of scale.

COMPETITIVE ADVANTAGE

Businesses are keen for disruptive technologies to become more competitive in a globalized economy that creates a great divide between winners and losers. Technological capabilities dictate cutting-edge business strategies that formulate the ability to win across a highly competitive landscape. The firms that grasp that concept are the ones that will rewrite the business rules for years to come.

One of the most important boosters of innovation is competition. It's easy to lag on a trend and not implement the latest technology. But this is not the case with AI. AI is much more than one technology. When talking about business, it is a combination of technologies that allow companies more predictability and accuracy without extra spending on a scale.

That's why no company can stand still and ignore the innovation and technology absorption. Doing that would cause an immediate failure or slow death. The speed of adaptation is also a huge factor. Competitive success predominantly favors those who can quickly coordinate and apply technological developments.

A perfect example of using competitive advantage is Cosabella, a lingerie retailer, replaced its digital agency with an AI platform. Since then, their ROI has more than tripled, and the customer base has increased by 30%.

Cosabella's Marketing Director, Courtney Connell, says the decision to switch the service provider came after going through a flat quarter right after enjoying steady double digit growth quarters before. After parting ways with the digital agency, she searched for alternatives and decided to try an AI platform instead of building up a dedicated in-house team.

Cosabella selected a platform created by Adgorithms called Albert. By the end of the first month, the platform optimized and decreased ad spend by 12% and increased return on ad spend by 50%. The challenge given to the AI-powered service was to identify and convert high-value audiences using the KPIs and creative provided by the brand.

The work of the automated platform in the fourth quarter of 2016 contributed strongly towards the company's 37% increase in overall website sessions, a 30% increase in new users, and 1,500 more transactions.

"After seeing Albert handle our paid search and social media marketing, I would never have a human do this again," said Cosabella CEO, Guido Campello.

This makes sense when you consider the difference in results the company gets for the same amount of money. But the fact is, it's not cheaper to use AI-powered services; it's simply more efficient. Among the strong points is the absence of down time, speed, and lack of human-related communication issues.

Another great example is the Japanese insurance firm Fukoku Mutual Life which is making 34 employees redundant and replacing them with IBM's Watson Explorer AI that can calculate payouts to policyholders.

The intelligence behind this AI is capable of reading information about thousands of medical certificates, length of hospital stays, medical histories and any surgical procedures to improve the process of calculating payouts. Though the time needed to formulate the appropriate calculations of payouts is drastically reduced

- the sums will still go through the verification and approval process by a member of staff before being paid.

Just think about how much more competitive this insurance becomes with a single decision to automate the whole department of payout calculations. A lot.

Though this firm still leaves the final decision-making part to humans, some companies are aiming to automate this layer as well since this type of automated decision-making provides businesses the precious time saving and elimination of human emotional volatility, which is often a target of manipulation in highly competitive marketplaces.

This is how the world's largest hedge fund, Bridgewater Associates, is building a software system to automate most of the day-to-day management that happens within the firm. This includes hiring, firing, and even strategic planning and decision-making.

The system is based on a 123-page public manifesto written by the founder of the firm known as the "Principles." The Wall Street Journal describes this set of principles as an "unorthodox management approach" involving "radical openness" that encourages employees to rate each other, openly disagree and criticize, and where honesty is a top priority.

What's fascinating is that Ray Dalio, founder of the firm, believes that people and organizations already work like machines, so much so that the word appears 84 times in his manifesto.

The proposed software, called PriOS, is designed to be able to predict meeting outcomes, autonomously mediate employee disagreement, and even dictate every single move of each employee's work day. The goal of the project is to overtake most of the daily management that happens across the firm's operations.

Bridgewater is already a highly data-driven company. Each meeting is recorded, and each employee is graded throughout the day using the so-called "dots" rating system. The company's intelligence lab has built another piece of software that incorporates these ratings into "baseball cards" that show employees their strength and weaknesses, including the ongoing tracking of the employee's goals.

PriOS also makes it possible to find the right staff for an open vacancy and even rule out the negotiations that happen across the team members when there are emotions involved or someone is having a bad day at work and their perception of the world is affecting the right decision.

This is a powerful example of how transformational changes across the organizational hierarchy may create a truly new competitiveness on the market. And we are going to hear more and more stories like this in the next few years just because it makes sense for businesses to pick services with artificial intelligence instead of plain old fully human operated agencies.

SYSTEMS OF INTELLIGENCE

Any sustainable and profitable business must have some defensive moats around itself. And in the recent and upcoming years, the number of dramatic shifts, including democratization of cloud processing and mobile, are rendering some industry standard moats useless and making it seem like it's almost impossible to build a defensible business. But that is not completely true and the concept of SOI (Systems of Intelligence) is exactly the type of a new defensible strategy any business can pursue.

The thesis behind this new moat is placing the new layer of intelligence across the existing stack of business systems that drive companies, such as systems of record (CRM, ERP, databases) and systems of engagement (web, mobile, chat), which allows companies to extract more valuable insights and create new type of value for the existing users.

These systems of intelligence seem to represent a new type of structure that combines the previously praised defensible strategies such as economies of scale (the notion of lowering operational costs with the growing demand) and network effects (the notion of the system having more value as more people use it). Systems of intelligence converge across the stack of existing infrastructure and combine the benefits of both scale and network effects through the intelligent usage of data, and it's growing supply.

The most prominent use case of systems of intelligence besides automation of existing workflows is the creation of new value-added workflows that haven't existed before and can be uncovered only with access to these existing systems of record and systems of engagement. Thus, the holy grail of business defensibility lies in the ability to create these new moats of existing infrastructure using machine intelligence.

BETTER UNDERSTANDING OF CUSTOMERS

There is so much data that we as a customer or a prospect leave everywhere in our digital footprint. All this data is impossible to analyze by human workers, but machine learning enables businesses to look at that data on a scale.

From segmenting your existing customers to lead scoring potential prospects, all this is becoming possible with a very tiny level of details that no human would ever notice. This is how Target figured out that one of their customers was pregnant.

Birth records are public, so as soon as a baby is born, new parents are inundated with special offers and incentives. Retailers compete to reach these parents first.

Marketers at Target wanted to market to women in their second trimester when they begin buying maternity clothes and prenatal vitamins.

The desire to collect information on customers is not new for Target or any other large retailer. You probably realize that every time you go shopping, you generate a lot of data and share even more data with each retailer. From your visits to online e-commerce storefronts of a big retailer to them further identifying you in the store. It may sound scary, but there are even indoor tracking technologies that can pinpoint foot traffic and its interests in various points of their shopping routine. And it would be a shame not to make a use of all this data for a business owner.

When possible, Target tracks the purchase history of their guests and keeps a record of everything they buy. The purchase history provides data that helps them uncover valuable knowledge about their customers. In this case, Target knew if they could identify expecting mothers in their second trimester, they could potentially market to them for years.

After a number of tests and analysis, marketers noticed trends in purchases that women made in their first 20 weeks of pregnancy, the beginning of the second trimester, and as their due dates approached. After going through the data, Target identified close to 25 products that when analyzed together allowed them to assign each shopper a "pregnancy prediction" score. This score enables

Target to send coupons at specific stages of a customer's pregnancy.

Months after Target began this strategy, a man walked into a Minneapolis Target and demanded to see the manager. He brought with him several coupons that were mailed to his daughter. He was unaware his daughter was pregnant and was angry thinking that Target was encouraging his high school child to get pregnant.

Another good example is how MOZ used AI to predict customer churn. They've designed a recurrent neural network to analyze user actions as a timeline and predict the future actions. Since actions customers are about to perform within the system are caused by a vast variety of factors from the past, it makes it possible to mine some valuable business insights and decrease churn of existing customers which has an enormous effect on overall company growth.

There are different factors that influence user behavior such as external factors (rules and regulations, changing economic circumstances, market demand), internal factors (change in income, urgent needs) as well as product related factors (previous experience with the product, the functionality required, the quality of the support). The ability to analyze all these factors with the respective historical timeline makes it possible to predict when the user will become a paying subscriber or will churn away from your business.

EXTENSIVE DATA ANALYSIS

One of the most popular words across the industries during the last big hype of "big data" was a search for insights. The major driver of that is obviously lots of data that businesses produce without extracting any knowledgeable value. That data is just being stacked on top of each other for years until someone decides to vanish it.

There is a reason why traditional statistics is failing to do its job in the era of such a vast data-producing environment. The goal is to find the relationship between certain causes and a certain outcome, based on the data that influenced it. In data sets containing thousands or hundreds of thousands of contributing data points, this type of statistical analysis is getting us nowhere.

But with machine learning, we can ignore some parts of the data by simply learning that it doesn't influence the general outcome that much. This process is self-evolving, meaning that some data may not influence decisions in one department but may influence in another and is a subject to context.

A good example is Microsoft where researchers attempted to build recommendation systems that scale promptly to accommodate the stream of new users visiting certain online services for the very first time, which is a long-battled issue of a cold start in data science.

They've used a novel user representation that allowed the deep learning model to learn relevant user behavior patterns and provide useful recommendations for users who have never interacted with the system with the assumption they have some existing search or browsing history on their profile.

This combination of data coming from sources of different domains allowed the creation of a new type of the machine learning model that helped improve the recommendation quality within both new and existing domains. Such deep learning architecture called multi-view deep learning was tested on three real-world recommendation systems built upon a range of Microsoft products such as Windows Apps recommendations, News recommendation engine, and Movie/TV recommendations.

The results of a proposed approach showed that the multi-view model is much better than the general algorithms with a 49% enhancement in recommendations on existing users. As for the results on new users, it increased the quality of recommendations by an impressive 115%. This type of new deep learning model can be easily scaled to support millions of users and databases with billions of entries, as per Microsoft analysis.

The results of this multi-view deep learning experiment have proven the concept that the combination of features from all domains produces much better performance in contrast to the build-out of separate models designated for one domain specifically. It even outperformed industry standard recommendation algorithms across several public benchmarks along with the company's internal dataset.

The results of the experiment are quite impressive since the recommendation engine was able to automatically connect the behavior of a user in one type of domain like searching for some news with the actions performed in an entirely different domain like searching for TV shows.

Even though the approach may seem trivial for a human, to link specific interests of a human with interests in another type of domain, it's still not the easiest thing to accomplish for machines, but with a proper architecture proves to be doable.

FASTER DATA ANALYSIS

Aside from getting an insight, there is also an important metric of "speed to insight" which puts you into another dimension of competition between who is going to benefit from the insight first.

As much as the speed of analysis relies on computational power, it is also dependent on the quality of the data and machine learning algorithms used. But if you assemble everything together you are set up for a tremendous return on investment.

For example, NASA has an AI system called Autonomous Sciencecraft Experiment (ASE) that recently ordered a satellite to capture images of a volcano in Ethiopia. The minute the volcano developed a fissure, NASA's spacecraft (EO-1) began taking pictures to document the event.

Since NASA ordered the photographs before anyone asked, ASE was able to notify researchers within 90 minutes of detecting the event and gave EO-1 new tasks within hours. This would typically take a ground team weeks to accomplish.

Another fascinating showcase of how the speed of insight is important happened at the hospital affiliated with the University of Tokyo's Institute of Medical Science in Japan where research team looked to IBM's Watson for a solution to the complicated case of leukemia.

After observing a myeloid leukemia patient have an unusually slow recovery from post-remission therapy, doctors concluded they were dealing with a different type of leukemia.

The doctors turned to IBM's Watson, a cloud-based, AI-powered supercomputer that is capable of cross-referencing and analyzing data from millions of oncology papers from all over the world and instantly pulls out the needed information, much faster than humans can. They cross-referenced the patient's genetic data with Watson's database and detected thousands of genetic mutations in her DNA and filtered out which ones were diagnostically important and not just hereditary characteristics unrelated to her disease.

Watson accomplished this feat in just ten minutes—human scientists would have taken two weeks. "We would have arrived at the same conclusion by manually going through the data, but Watson's speed is crucial in the treatment of leukemia, which progresses rapidly and can cause complications," Professor Arinobu Tojo, who led the research team, was quoted by The Japan Times as saying. "It might be an exaggeration to say AI saved her life, but it surely gave us the data we needed in an extremely speedy fashion."

Watson's quick work helped the researchers to find out that the patient had a rare secondary leukemia caused by myelodysplastic syndromes, a group of diseases in which the bone marrow makes too few healthy blood cells.

IMPROVED EFFICIENCY OF INTERNAL PROCESSES

From improved manufacturing to cloud server operations, artificial intelligence offers unseen benefits. By adopting and restructuring your business processes around AI, you are set to acting as a machine learning system yourself. Since machine learning is actively optimizing the end output, you can optimize your internal operations to support that exact end goal.

Like Bridgewater Associates, the firm that tried to build itself as a machine with a set of rules run by AI, there are other firms aiming at restructuring themselves around machine automation.

In 2000, Goldman Sachs's headquarters in New York employed over 600 traders that were buying and selling stock on the orders of the investment bank's large clients.

Today, there are just two equity traders and the rest of the work has been successfully automated and is supported by 200 software developers.

Complex machine-learning-powered trading algorithms are coming for more sophisticated trades after replacing operations where the price of the sold assets was determined through some easy-to-follow logic. This means things like currencies and credit that are traded through less transparent networks of traders are being targeted for automation. These operations are powered by algorithms that are learning to behave like human traders do, but much more time efficient and with fewer errors.

As much as internal processes are automated, the external communication is not left untouched either. Research firm Gartner has published predictions that by the year 2020, most customers will experience customer service without having to speak to a human being.

This means that the systems and automated inquiry handling systems that are currently dealing with these requests will soon be upgraded by a more advanced type of machine intelligence.

Working with AI as a part of improving internal processes has a learning curve to navigate. It requires a certain mindset change.

ENHANCED R&D CAPABILITIES

In addition to improving existing processes and gaining a better understanding of your customer base, AI enables companies to create new types of products that wouldn't be possible to create without the AI component in it.

A great example is Replika, the app to mimic a user's personality which was started by Eugenia Kuyda as a side project, a simulacrum to bring her best friend back to life.

Eugenia lived in San Francisco with her best friend, Roman. One day, Roman was killed in an accident while walking across the street.

For months, Eugenia read through text messages from Roman. She felt they had a unique connection and didn't have anyone else to have those types of conversations with. Then she realized she had the technology to create chatbots that could solve this problem.

Eugenia created a chatbot that would text her and other friends to text the same way as Roman did from the text message history between her and Roman. She could communicate with him like he was still there.

They talked about changes in their lives, told him they missed him and heard what he had to say back.

Replika then became a larger-scale experiment to mimic living people. It's a personal chatbot you can improve by texting with it. By chatting, you teach it your personality and gain points. It can also chat with your friends and learn more about you from other people. The plan is potentially to create a bot to take over some of the routine messaging responsibilities like social media, or checking in with your family.

The team behind Replika have also created Sessions, an AI-powered journaling feature that offers people all the benefits of self-journaling without the extra stress. You tell the chatbot how your day went, your accomplishments and roadblocks, your dreams and thoughts. The bot captures all this information and stores it as a journal.

The concept comes from the fact that conversations about your day with others come naturally easy while writing your thoughts on paper can be intimidating. You can always find things to talk about with your friends after a hard day because you know that they care.

Replika asks questions that matter to you, remembers what you've said, and helps you make sense out of it.

AI will enter the workplace with this type of innovational products affecting what we do. Digital assistants will soon be the primary way people interact with their smart devices. Samsung's new digital assistant, Bixby, goes beyond answering basic questions like today's weather and helps users control the phone. Bixby responds to voice commands and completes tasks, such as finding a photo and sending it to another person.

"When the smartphone came out, touch interface became the norm," head of R&D for mobile software and services Injong Rhee said. "Ten years after the introduction of smartphones, another revolution is waiting. That revolution comes from machine learning and deep learning."

COST REDUCTION

Human time is expensive, but you can reduce that expense when you can maximize the time spent on the right things instead of repetitive routines. You can also save on other resources, such as spending less money acquiring 3rd party intelligence or having fewer seats on an online service that allows you to do your job better. Or even saving on energy.

Costs on your existing infrastructure can be cut just by optimizing it using machine learning as Google did. Google saves hundreds of millions of dollars on power consumed by its data centers by using technology from the DeepMind artificial intelligence subsidiary.

Much of the energy used in the data centers is to provide cooling for the many servers running. Large industrial machines are used to cool the servers, but it's difficult to optimally optimize the operations because traditional rule-based engineering and human intuition can't capture the interactions between the data center environment and the super complex equipment.

Also, systems can't easily adapt to external changes around them like weather or meteorological changes. There are too many scenarios possible and different environmental factors to program rules for each one.

Each data center has unique architecture and environment which makes the custom-adapted, specific-to-environment models useless since they can't be moved and applied equally effective in another setting.

DeepMind researchers (previously famous for teaching AI to become pro player of Atari games) began using a system of neural networks trained on different operating scenarios and parameters within the data centers.

They created a more efficient and adaptive framework to understand data center dynamics and optimize efficiency by taking the historical data from thousands of sensors within the data center – data such as temperatures, power, pump speeds, set points, etc. – and using it to train an ensemble of deep neural networks.

So, just as machine learning algorithms follow the game play and make moves in Atari games, smart algorithms control more than 120 variables inside the data center environment to "win" and get the highest score in a resource optimization game. The game rules are obviously different, and the highest score is a more efficient consumption of electricity through the control of cooling systems, windows and other things.

MACHINE LEARNING DATA CENTER POWER CONTROL

HIGH POWER USAGE

ML CONTROL ON

ML CONTROL OFF

LOW POWER USAGE

Source: Google DeepMind

Google described the 40 percent reduction of electricity needed for cooling as a "phenomenal step forward." This reduction translated into a 15 percent reduction in overall power, saving Google hundreds of millions of dollars over the years.

Optimizing energy consumption has become popular over the last few years. New companies such as EcoIsMe assist homeowners in reducing their bills and companies like Verdigris help large organizations optimize consumption by using data collected from large facilities and automatically detecting energy-consuming devices based on an individual electronic fingerprint.

The energy sector is not the only heavy industry where AI is gaining momentum. Machine learning's core values align well with the complex problems manufacturers face daily. From striving to keep supply chains operating efficiently to producing customized, built-to-order products on time, machine learning algorithms have the potential to bring greater predictive accuracy to every phase of production.

Many of the algorithms being developed are iterative and designed to learn continually and seek optimized outcomes. These algorithms iterate in milliseconds, enabling manufacturers to seek optimized outcomes in minutes versus months.

Manufacturers often are challenged with making product and service quality to the workflow level a core part of their companies. Often, quality is isolated. Machine learning is revolutionizing product and service quality by determining which internal processes, workflows, and factors contribute most and least to quality objectives being met.

Using machine learning, manufacturers attain much greater manufacturing intelligence by predicting how their quality and sourcing decisions contribute to greater performance.

Smart manufacturing systems designed to capitalize on predictive data analytics and machine learning have the potential to improve yield rates at the machine, production cell, and plant levels.

Machine learning is making a difference on the shop floor daily in aerospace and defense, discrete, industrial and high-tech manufacturers today. Manufacturers are turning to more complex, customized products to use more of their production capacity, and machine learning help to optimize the best possible selection of machines, trained staffs, and suppliers.

IMPROVED EFFICIENCY OF EXTERNAL PROCESSES

For many complex manufacturers, over 70% of their products are sourced from suppliers that are making trade-offs of which buyer they will fulfill orders for first. Using machine learning, buyers and suppliers could collaborate more effectively and reduce stock-outs, improve forecast accuracy, and meet or beat more customer delivery dates.

The evolution of using AI and machines that learn in supply chain planning is coming. There are already early examples of the potential of AI to improve both supply chain planner efficiencies and provide better or optimized supply chain decisions.

A good example is SAP Ariba, which unveiled an AI-powered assistant named *Procurement Bot* to speak with buyers and suppliers about orders in a conversational interface. The company wants to make issues like errors in invoices and various miscommunications a thing of the past.

SAP Ariba plans to give both sides of the buyer-seller marketplace ability to talk with their bot as they would talk to Siri or Alexa. Powered by machine learning algorithms, the bot will be able to inspect and learn a company's policies and preferences regarding different procedures. Combining these individual needs with guided recommendations, it will help reduce the human error and increase the speed of operations across the marketplace.

Machine intelligence is also changing how enterprises optimize their pricing routines. Much of the focus lies in the ability to optimize the pricing alongside with recommended strategies to improve sales cycles and bring in more opportunities.

Inventory position optimization is a great example how machine intelligence can be utilized to automate supply chain routines. One of the causes of the extremely high inventory-to-sales ratios in the US right now is an emergence of omnichannel retailing. The free and fast shipping and easy returns alongside the everyday low prices has forever changed how customers buy things and how the products reach the end consumer. This "Amazon effect" has caused a real struggle to keep up with the change that forces companies to juggle with the inventory positions across different locations.

Retailers can automate the process of inventory buffering at the right locations and provide optimal inventory positions by embracing the use of machine intelligence. Total inventories can be reduced by almost 30% while still matching the demand.

The right AI strategy can prove to be very beneficial for the supply chain optimization. Even a gradual introduction of machine learning can dramatically improve the current demand forecasting using correct prediction models.

This would also transform how planning departments work in the coming years.

HUMAN ASSISTANCE

As much as AI will transform the future of work, it will create a land of opportunities for the new type of machine-human cooperation. Long gone are the days where you use static tools to help solve some specific complex equation or situation.

Machines are going to take off a load of mundane routines for all levels of workers across many industries. Can you imagine that just hundred years ago more than half of our population was involved in agriculture and making sure we have enough food produced? Now it's less than a few percent. And that's the ultimate business goal, to make humans more productive and prevent them from doing things that can be automated.

Human assistance is a hot topic as we see more and more machine augmented processes. Companies like Digitalgenius are on track to automate customer support and what they do is exactly the right usage of humans.

The company recently released a new product called the Human+AI Customer Service Platform that integrates with popular customer service platforms like Salesforce, Zendesk, and others. The experience of the product is quite similar to the chatbots that are gaining much attention lately. While chatbots sound cool in theory, they often break once facing the reality of customer support when users ask for something that has not been pre-defined as a command into the knowledge base.

The combination of humans and machine intelligence that DigitalGenius offers is creating a new way of doing customer service. This seamless experience is not limited by the pre-defined actions or phrases which bots usually use. Machine learning capabilities of the platform provide a way to analyze the customer service logs from the previous conversations with representatives and learn about the most common transactions with the respective most likely responses.

The concept of a "confidence threshold" defines the likelihood the response is appropriate for the specific question. Customer service representatives make an educated decision on whether the automated reply is good and can change it or reply themselves if needed.

It is a great use case of machine intelligence serving for human assistance and one of the most practical uses of machine intelligence when it comes to customer service.

TAXATION

A recent Quartz interview with Bill Gates has unleashed a heated debate on whether governments should tax robots and other advanced technology which replaces human labor. The launcher of software behemoth Microsoft and one of the richest people on earth advocated taxing robots as a way of slowing the adoption of automation and redirecting employment.

"Right now, the human worker who does, say, $50,000 worth of work in a factory, that income is taxed, and you get income tax, social security tax, all those things. If a robot comes in to do the same thing, you'd think that we'd tax the robot at a similar level," argued Gates. He cited the perennial need for tax revenue to fund elderly care, reduce school classes, and train new workers.

The furor Gates caused aside, the idea of taxing machines is far from new. Governments around the world have so far been reluctant to tax manufacturing or farming machines. European Union legislators debated and ultimately rejected such a proposal as recently as last February. The revenue raised from the robot tax was intended to go into training programs for workers outplaced by automation.

The topic remains live on both sides of the Atlantic. In a recent interview, US Treasury Secretary Steve Mnuchin said it was unrealistic to expect AI to supplant human labor in the next fifty to a hundred years, adding he was optimistic about automation.

Many, including advocates of automation and robotization, fear that robots and AI-powered manufacturing would destroy human jobs and increase inequality. Others argue that robots and AI systems are part of an overall technological advance that has created a veritable mushrooming of work opportunities since the Industrial Revolution. At stake in the debate is whether automation is viewed a danger or an opportunity.

In a sense, the current debate's precursor has been hibernating in social consciousness ever since the Industrial Revolution. That ancestor of the age of AI saw a political movement, the Luddites, passionately opposed to what they saw as machinery destroying the traditional fabric of their *milieu* and depriving it of livelihood.

Governments today could do worse than recall how things unfolded back then. The Luddites are a footnote in dusty history textbooks, while the mechanical windmills at which they charged Quixotically form the acknowledged foundation of today's industrial society.

For the time being, taxes on robots appear to have been deferred. Nevertheless, clamor to introduce them is certain to return.

Another aspect of machines, and one which is often underrated is that most of them are surprisingly flexible. Reprogramming and retooling robotized production lines is generally easy and quick. Add flexibility to uniform manufacturing quality, and businesses can plan in a way that was inconceivable in the era of manual production.

Modern robots are complex, complete systems combining hardware, software, sensors, and controls. They offer their proprietors much greater control over feedstocks and manufactured output, down to the last component. They can be tweaked to fit changing production cycles.

While the archetypal symbol of progress in the early 20th Century was Henry Ford's conveyor line, AI systems connected over the Internet of Things have replaced that in the first two decades of the 21st Century.

Returning to the vexed issue of whether to tax machine intelligence, the debate will probably reach a robust and loud peak over the coming decades.

As things stand, legislators might well reaffirm the time-worn adage that if something is pleasurable or beneficial, it ought to be taxed. Manufacturing and farming should make themselves heard in the debate with a united voice. Briefly put, taxing machines can harm entire industries and damage the future of innovation, to the extent of jeopardizing the world's entire economy.

Over the past 30 and more years, a significant chunk of world manufacturing has moved to China. Growth has been stagnant since the 2007/2008 crisis that has failed to produce the rebound that followed prior crises. World banking is practically on artificial respiration. Against this background, the last thing we need is a tax on machines. Indeed, MIT economists, among others, have argued that the current pace of innovation is far too slow and that it is only innovation that can pull the world economy out of its current quagmire.

Logic dictates that businesses should introduce more machines to boost productivity and effectiveness. But would they bother to do so if they knew their machines would be taxed? And where would refusal to adopt robots lead to a spiral of reduced research and development and consequent productivity and competitiveness drops?

There is another dimension to the debate. The world's population is set to grow fast. Billions need to be fed: something getting increasingly tenuous without injecting new technologies and automation into agriculture. Why slide back to the Middle Ages when farmers can happily keep feeding the world – provided they adopt advanced technology?

Add to that a workforce shortage. The baby boom generation is getting ready to leave the scene. Statistics show that 76 million members will retire within the next thirty years. Only 46 million new workers will be available to replace them in the economy. That is a net workforce loss of 30 million or almost 40 percent. The same trend applies to developing economies as well. Urbanization is also a factor. People are leaving rural areas so fast that even traditionally rural China is reporting rural workforce shortages.

Sustaining innovation is crucial for global economic growth. Businesses should be reassured that their innovation moves will not be penalized by taxation. Governments should seek alternative methods of handling the social and political effects of AI, automation and robotics.

Machines will encroach on many traditional human endeavors in the very near future. This is as inevitable as the typists and typing pools becoming obsolete in the early computer and software revolution which was barely a generation ago. The question legislators and business leaders should be pondering is how to outplace the human employees in a humane and constructive manner, rather than how much tax to levy on robots.

Finland's government, for example, was the first to launch a national basic income program. This is currently a pilot scheme, pending a decision on whether to extend it to all Finns. A varied group of unemployed people each receive a $583 tax-free monthly stipend and will continue receiving it even if they were to find a job. Money for the program is not raised by hiking taxes but skimmed painlessly from a highly effective economy led by innovative companies like Nokia. It falls off the back of innovation and smart spending not by burdening innovative companies.

Past technologies mainly complemented human labor. We cannot take it for granted, however, that evolving technology and a technology-driven society will continue working the same way. Medieval society was dominated by clear class divisions, mostly based on lineage and trades. The 19th and 20th centuries saw the rise of the industrial society, with monarchies disappearing, workers moving into suburban homes, and traditional businesses adopting modern machinery. Robotics are ushering a new economy that is the next inevitable phase of progress. How labor will adapt to these new conditions is difficult to foresee. One thing is not difficult to foresee, however; there is no alternative way for the global economy to grow.

A PwC report hints that automation might be a source of increasing wealth and unsuspected new employment.

Their analysts believe that robotics and AI technologies would create several brand new digital technology occupations. Productivity gains would generate more wealth and spending. In turn, they would support more job creation, mainly in service industries where automation is harder to implement and service must be human. PwC claim that some six percent of UK jobs in 2013 were in occupations that had not existed before 1990. Most were related to digital technologies. By 2030, the report continues, one in 20 UK employees would be working in fields like AI and robotics. This would largely offset the much-feared implications of extensive automation and robotization.

The machine taxing debate appears premature, if not hyperbolized. Businesses must invest in automation and research and development efforts in this field should be encouraged not restricted. Governments ought to be made to realize that taxing automated systems would not solve any problem. They ought also to acknowledge that progress follows arcane rules not subject to étatist fiat.

KNOWLEDGE PRESERVATION

Preservation of acquired knowledge has a long history, from cave paintings to books and wikis. This knowledge preservation development has gradually transformed in the 21st century with the emergence of computation and information management, which led to the creation of knowledge management tools. Machine learning has the potential to innovate how knowledge is created, organized and used within organizations.

AI can provide a way to prevent the knowledge from being lost when the key people are no longer available to the organization, especially in an organization where the knowledge is tied to one person or a group of people.

Knowledge management personnel already use machine learning tools to create, organize, clean and reuse knowledge within organizations. Machine intelligence provides companies with tools to analyze and classify texts, automate reasoning and visualize information to improve decision-making.

This way companies like DigitalGenius are using existing knowledge storing capacities to produce value out of the data. These companies also provide the tools to process raw human input such as voice recognition and handwriting. Companies like Tetra and others use natural language processing to record your phone calls and transcribe them into text for you to easily search and remember everything discussed.

The beauty of the knowledge preservation is that this knowledge can exist if it is needed and is relevant to the types of problems companies solve.

Moreover, machine intelligence systems help to improve knowledge management and provide the augmented infrastructure that helps to use expert systems more efficiently.

These machine intelligence systems are able to operate huge volumes of data and even introduce new ways to store this data through the use of decentralized systems such as blockchain. Thus, various organizations promote artificial intelligence to boost and optimize the routines of knowledge workers.

LIMITATIONS

Besides having plenty of benefits, there are still plenty of limitations that businesses should be aware when diving into the world of AI.

One such limitation is zero-day issues, the things that never happened before are entirely outside of the scope of machine intelligence. Even rare, not unique events still present high limitations for machine learning algorithms because it means there is just not enough data for AI to work with, build models and predictions.

Though zero-day issues are common and almost untreatable by the usual machine learning means, it is important to catch them, the sooner - the better. Even though this is a limitation, some companies are using it as a core part of the business by optimizing algorithms to detect those anomalies.

As much as machine intelligence is susceptible to the absence of data, it is also does not perform well with the presence of bad data. Bad data can be a result of human error or the specifics of operations inside the company. Nonetheless, it will affect algorithms in a negative way by directing it towards other decisions which may not always be traceable and could cause serious financial trouble.

This inability to trace the decision back to its cause leads to another limitation. More commonly now, with the advent of deep neural structures, creators of machine learning intelligence can't really explain how it works. This black box effect is mostly the product of the neural network architecture that is not built to be explainable.

The systems built are too complicated for its developers and researchers to find the exact reason for any decision that is made. And there are no simple ways to question the system since current machine learning models are quite limited and have no internal ability to explain why specific action was taken.

A perfect example is Google's newest RankBrain search results engine. Unique updates to the search algorithm are helping the system to understand more ambiguous queries better. Machine intelligence makes its best attempt to guess what your query is related to even if it's phrased poorly, similar to how humans can easily answer those types of questions. Google engineers reportedly admitted that they have no idea how it works inside, simply because of how complicated it is.

So, if you ask why certain search result shows up in your results, it would be almost impossible to trace back why. Recommender systems that are already in place and used in banks, hospitals, cars, military and hedge funds, have a similar problem.

There are already plenty of arguments that demand an ability to understand why the system has reached a certain conclusion and claim it as a fundamental legal right. These explanations regarding machine decisions may become a required feature starting in 2018 per "right to explanation" proposed by the European Union.

This creates a huge problem for even relatively simple systems like movie or song recommendation apps that use machine intelligence to predict what's the best movie for the user to watch next. So, we can only imagine how crazy it will get for complex systems such as DeepPatient which predicts the future of patients based on the available health history and patient records.

DeepPatient is built using multiple layers of neurons with more than five hundred hidden neurons per layer, each layer and neuron is interconnected and holds its abstraction, which makes it impossible for humans to even dive in and try to analyze why the final prediction is what it is.

Electronic Health Records

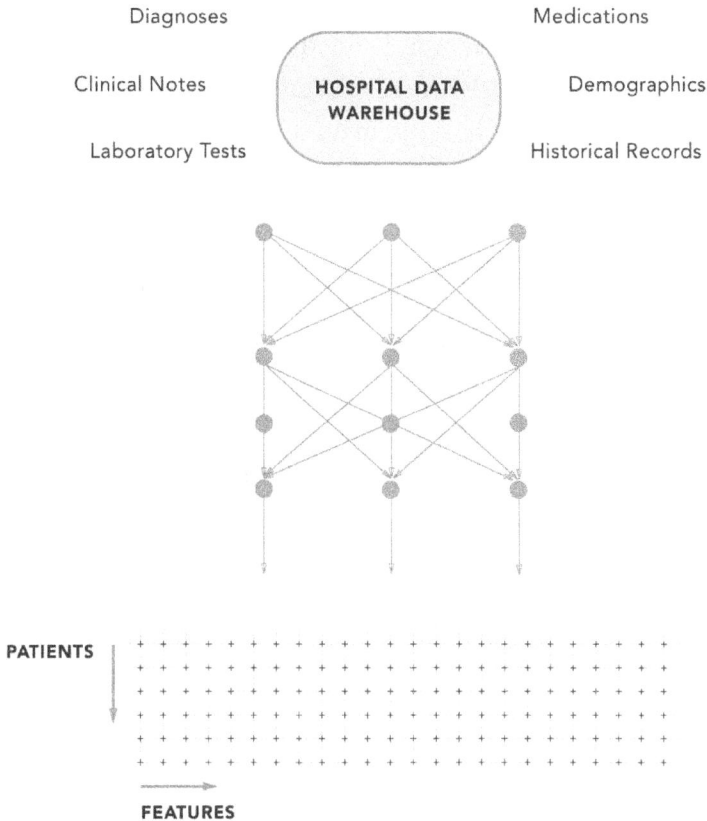

Diagnoses

Medications

Clinical Notes

HOSPITAL DATA WAREHOUSE

Demographics

Laboratory Tests

Historical Records

PATIENTS

FEATURES

The system itself proves to be super-efficient in predicting such diseases as diabetes, schizophrenia and even some types of cancers just by analyzing current health records of a patient. Though no doctor can tell why it comes to such conclusion, which is quite worrying considering the life-threatening nature of the application.

There were several attempts to tackle this issue. One of them involved generating images out of those trained neural nets. Google researchers experimented with an image recognition deep neural network to generate sample pictures of what it was trained on instead of spotting those objects. The results of running the algorithm in reverse allowed the system to visualize those features that neural nets identify. The resulting project called *Deep Dream* that has harnessed plenty of press was generating alien-like animal pictures emerging from dogs and plants, a very hallucinatory experience to be honest. Even though the Deep Dream experiment showed that algorithms do highlight some of the features we understand like bird's feathers or cat eyes, it also proved how different neural net "perception" is from what we are used to.

These limitations also proved to be a way to hack machine learning when a group of researchers showed how certain manually generated images could fool deep neural network into perceiving things that aren't there, just because the images are generated in the way to exploit these low-level patterns the system searches for.

Defense Advanced Research Projects Agency (DARPA) is one of the companies that are very interested in the ability to interpret the results from machine learning systems. It's no surprise that the military is funneling billions of dollars into projects that would pilot vehicles and aircraft, identify targets and even eliminate them on autopilot. That's why here it's even more important, maybe even more than in any other field, to be mindful of the results AI presents.

Explainable AI, the initiative funded by DARPA is called an essential measure if future soldiers must understand, and most importantly trust, machines to effectively manage this way of partnering with AI.

EXPLAINABLE ARTIFICIAL INTELLIGENCE

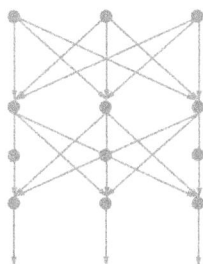

CAT

CURRENT EXPLANATION

THIS IS A CAT

XAI EXPLANATION

THIS IS A CAT BECAUSE
IT HAS WHISKERS
IT HAS FUR
IT HAS FELINE EARS

XAI program goal is to create a set of new machine intelligence techniques that:

- Create more explainable machine learning models, while keeping a high level of accuracy and efficiency

- Enable human operators to understand, trust, and effectively manage the use of AI across industries

These new machine learning techniques will enable machines to explain their decisions and describe which things influence its successful operation to trust and use them more predictably in the future.

TODAY

| TRAINING DATA | → | MACHINE LEARNING PROCESS | → | LEARNED FUNCTION | DECISION → | USER |

WHY DID YOU DO THAT?
WHY NOT SOMETHING ELSE?
WHEN DO YOU SUCCEED?
WHEN DO YOU FAIL?
WHEN CAN I TRUST YOU?
HOW DO I CORRECT AN ERROR?

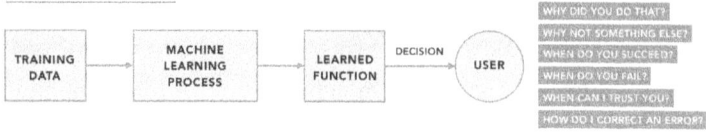

EXPLAINABLE AI

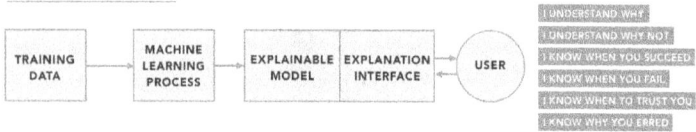

| TRAINING DATA | → | MACHINE LEARNING PROCESS | → | EXPLAINABLE MODEL | EXPLANATION INTERFACE | → ← | USER |

I UNDERSTAND WHY
I UNDERSTAND WHY NOT
I KNOW WHEN YOU SUCCEED
I KNOW WHEN YOU FAIL
I KNOW WHEN TO TRUST YOU
I KNOW WHY YOU ERRED

These models will require new types of human-computer interfaces that can translate these complex models into understandable and useful explanation dialogues for the end user of the system.

Bias is also a strong factor hurting the AI algorithms. Even though it's quite a known limitation, researchers and companies are only now feeling the real effects of it, when the scale reached its full potential.

A well-exaggerated example is how a robot board of judges decided the outcomes of the International Beauty Contest. The promise of the contest was to showcase that algorithms can precisely assess the criteria connected to the objective evaluation of human beauty. Due to the biased training dataset, the algorithm failed with having only white contest winners.

This is an example of so-called data-driven bias, which is a well-acknowledged industry term but sometimes gets ignored assuming that huge datasets are not skewed, which produces skewed results. The same way Nikon camera became Internet famous for detecting "someone blinking" when pointed at the Asian faces or HP computers were proclaimed racist for not detecting black faces on the camera.

Overall, it's important to understand limitations hidden within the numerous advantages and no magic wand exists that would make all problems go away and make the company more successful.

CHAPTER THREE

WHICH AREAS WILL BE AFFECTED THE MOST?

This section covers the areas that most likely will be disrupted by AI advances

AUTOMOTIVE

AI has the power to radically transform the automotive industry in the coming decades. Self-driving cars have been successfully tested since 1995 (when the first coast-to-coast drive from Pittsburgh, PA to San Diego, CA was completed), and are becoming more sophisticated and more intelligent. Governments are quickly adapting legislation to the requirements of new technologies, and old automotive giants are developing solutions for their cars.

Although broad adoption of fully autonomous cars might not come anytime soon, major automotive innovators such as Tesla are integrating various semi-automated and fully-automated solutions into their conventional car offerings now. AI revolution parallels the growing digitalization and connectivity of cars, which has a disruptive impact on the entire concept of mobility and car ownership. AI expansion into the automotive industry means that by 2025, 100 percent of new automobiles will be connected, and by 2030, 15 percent of new cars will be fully autonomous.

The emergence of the AI-enabled automotive market brings both opportunities and challenges for automotive companies. Most old automotive businesses depend on the model of private car ownership. This model, however, is likely to be challenged by the growing popularity of shared mobility. With more self-driving cars for sale, more consumers will be attracted to car sharing rather than private car ownership. On the other hand, self-driving cars that can work 24/7 without any human input will have a shorter service life. As a result, automakers will witness a surge in new orders from shared mobility services and transportation companies that will offset the decline in private orders. The rapid growth of the self-driving car market may also be facilitated by the enhancement of road capacity generated by vehicle-to-vehicle communication integrated into autonomous vehicles.

Digitally connected self-driving cars can exchange road traffic information to prevent collisions and to minimize traffic jams. With 100% of vehicles connected in the US, the road capacity can reach 12,000 passenger vehicles per hour, up 445% from 2,200 pc/h per lane we have today. Mobility opportunities for children, elderly, and disabled passengers will also boost the high demand for self-driving cars. These population groups will directly benefit from mobility-as-a-service opportunities offered by self-driving cars.

AI Solutions in Contemporary Cars

Self-driving cars are not necessarily the only manifestation of the AI revolution in the automotive industry. AI features and software are now increasingly integrated into conventional cars to offer a wide variety of semi-automated solutions, digital services, and entertainment features to drivers and passengers alike.

For example, major car manufacturers are already producing various Advanced Driver-Assistance Systems (ADAS) developed to automate and adapt vehicles for safety and better driving. Along with automatic braking and cruise control systems that have been around for many years, contemporary AI-powered ADAS include cross-traffic alerts, adaptive light controls, automatic parking, automotive night vision, traffic sign recognition, and emergency driver assistance.

For example, advanced semi-autonomous systems have been offered by Tesla in its *Autopilot* feature that allows Tesla cars to accelerate, change lanes, and move from parking spaces and garages autonomously. Another emerging trend is the introduction of complex infotainment systems that allow for interactive control of car interfaces via speech recognition and computer vision technologies. Owners of such cars as the 2015 BMW 7 Series can easily manage these via voice commands interpreted by embedded wireless machine learning systems.

Mentioned innovations turn automakers into de facto tech companies that put effort into the production of the state-of-the-art software packages and features in their car offerings. AI-powered software systems, infotainment, and other car control interfaces expand the range of car services that may be offered to clients as premium features and special packages. By integrating these software packages and features into automobile electronic systems, automakers substantially expand the scope of their investment and profit opportunities and sources.

Marrying AI, Big Data, and Cloud Technologies in Cars

AI technology in the automotive industry is also propelled by the rapid development of big data and cloud computing. Machine learning algorithms use big data retrieved from cloud databases to learn consumer preferences, behaviors, and needs, to create customized solutions for drivers and passengers. Cloud-based intelligence platforms learning from big data have been recently developed by General

Motors (GM) in partnership with IBM supercomputer Watson, a powerful AI system capable of answering the most difficult queries and questions using its state-of-the-art data mining and learning algorithms. An updated version of GM intelligence system, *OnStar Go,* now can produce useful suggestions of products and services in proximity to the car, or inside the car by analyzing data generated by drivers and passengers.

AI systems powered by big data and consumer analytics can help locate gas stations and pay for fuel from inside the car, identify nearby restaurants like those preferred by the driver, remind drivers to buy various household items on the way to relevant stores, or make video and music suggestions depending on the driver's aesthetic judgment.

In themselves, AI systems such as *OnStar Go*, contribute to deeper integration of automotive industry, banking, cloud services (e.g., Amazon AWS, Google Cloud), e-commerce, service sector (restaurants, hotels), retail, and entertainment into one service bundle delivered to consumers via data-driven recommendation algorithms similar to mobile personal assistants.

All these services are unified under the one umbrella of AI-enabled car systems. As a result, automakers and dealers enjoy better exposure to diverse markets and sectors of the economy taking part in the distribution of revenue flows generated by them. Ipso facto, the use of AI systems in cars has an enormous potential for the entire economy. According to the international consulting company Accenture, the total business value of connected car services is going to reach €100 billion by 2020, and €500 billion by 2025. Each car equipped with AI systems can generate additional €5,000 of value throughout its lifetime.

AI and Internet of Things (IoT)

AI systems in cars may also be paired with Internet of Things (IoT) infrastructure. IoT refers to the network of Internet-connected devices (sensors, timers, engines) that send information to cloud-based servers. This data is then processed and analyzed by powerful machine learning algorithms to generate useful insights about the performance and security of devices.

Internet-connected cars effectively become a part of this booming IoT market with a projected volume of $267 billion by 2020. Car data piped via IoT and analyzed by in-house AI systems of automakers is becoming a powerful instrument of automation, service customization, and fleet management in modern automotive industry. For example, in a cognitive predictive maintenance approach promoted by such companies as *DataRPM* software leverages car connectivity to predict potential problems in engines or braking systems and programmatically fix them on the fly avoiding costly visits to repair shops.

Via over-the-air (OTA) systems, car manufacturers, and dealers can instantly send car software updates and security fixes, or various premium features, to the vehicle's firmware. These updates and features will encourage car owners to spend more money on their cars in a variety of ways not available before AI and IoT became available. According to Navigant Research, collectively OTA updates and premium features in cars have a potential to add up to $21,7 billion annually in alternative services profits by 2023.

Additionally, IoT infrastructure powered by AI may be used to analyze performance and efficiency of cars to improve car design, ergonomics, and other parameters. For example, AI-powered vehicles can directly report fuel usage and refueling costs to fleet managers who can use this data to enhance the fuel efficiency of their car models. Similarly, smart biometrics sensors can be successfully used to enhance driving security. For example, via eye tracking sensors, car firmware can predict if a driver is tired or distracted, and send corresponding alerts to prevent emergency situations.

A combination of AI technology, car connectivity, IoT, and big data transforms the automotive industry into a testbed of innovation. Use cases of AI in cars are infinitely diverse, as well as its potential to create new products, services, and business models for car manufacturers and other businesses.

At the same time, however, digitalization of cars introduces new security risks and challenges for automakers. In 2015, Fiat Chrysler had to recall 1.4 million cars after hackers managed to take control over Jeep Cherokee via its Internet-connected entertainment system. Hackers could remotely control Jeep's engine and braking system whenever they wanted.

In addition to hacker attacks, AI technology in the automotive industry faces other complex challenges, such as ethical decision-making in self-driving cars, the reluctance of individuals to forfeit control of their cars, or using AI systems as weapons of terrorist attacks and other dangerous crimes. Notwithstanding these risks and challenges, advantages of AI for automakers and other businesses far outweigh the risks involved. AI capacity to generate additional services, features and value propositions by car manufacturers turns this technology into an excellent investment opportunity for automotive businesses and the wider economy.

AGRICULTURE

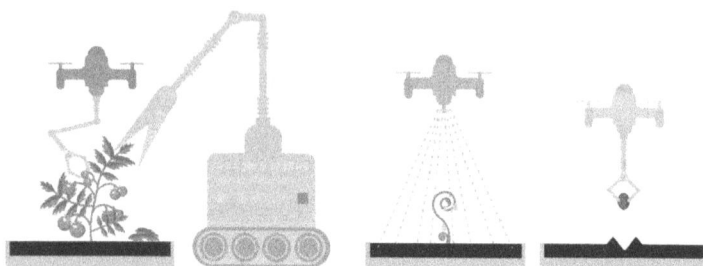

Global agriculture is a $5 trillion industry. A report by IBM shows it accounts for a tenth of global consumer spending, two-fifths of employment, and one third of greenhouse gas emissions. It is a traditional industry where attitudes and heritage play a crucial part. It's also subject to government attention to an extent unimaginable in any other business. Agriculture is also arguably the largest industry in which automation, robots, and AI are still to make serious inroads.

Demand for food will continue rising steadily and stably. By 2050, the world's population will grow to 10 billion. This makes the need to produce more food in smarter ways pressing. AI and robots offer the most convincing promise when it comes to reshaping agriculture for sustainable output growth.

Many attempts to introduce innovative technologies and AI in agriculture and farming have failed because of meager funding, insufficient testing, or resistance to innovation in what tends to be a family sector dominated by tradition.

Despite ingrained practice, today's farmers are beginning to use remote sensors to monitor their produce or check on farm and field conditions. Far more powerful applications are in development. Thus, using robots to tend crops is an old idea. Projects for farming robots are now in progress. Current thinking is that such robots would be deployed in networked fleets.

Alongside replacing human muscle in field work, mobile robots would also collect data which human farmers can easily overlook, or whose importance they might fail to perceive. Plant geometry, the ratio between leaves and fruit, are examples of barely perceptible indicators that repay precise periodic measurement. Once the robots collect this data (and others like it), they would send it to computers for analysis. These analyses might prompt seemingly minor adjustments to irrigation and fertilizing cycles, among others, to boost long term or seasonal yields.

The emergent Internet of Things (IoT) has attracted growing attention over the past decade. Its headline applications have tended to focus on urban and domestic facilities. The real promise of IoT is in manufacturing and farming. IoT technologies promise multiple ways to help farmers monitor soil, seed, and livestock, crop, water and fertilizer conditions, scrutinize equipment utilization, and control costs.

Leading software and hardware vendors have worked for some years on machine learning technologies for the agriculture industry. These would allow machines to perform real world tasks without human intervention. Cognitive IoT platforms can help farmers process vast data streams and stores. Such platforms could collect data from remote mobile sensors, compare the information with historical conditions, and then provide insights and suggestions on how to improve yield.

IoT research and development is addressing the ways in which mobile machines, like agricultural drones, may scan fields. They would provide AI platforms with real time data. Farmers would then take timely actions on areas of concern.

AI would also help farmers determine the exact soils suitable for particular crops, or match seed stocks to available soils to boost yields while avoiding soil exhaustion.

Combinations of fixed and mobile sensors can collect historical and current soil and weather data and forward it to AI platforms for analysis.

AI can also improve farming performance by automating tasks and enabling farmers to work remotely. The UN World Urbanization Prospects Report estimates that two-thirds of the world's population will be urban in 2050. The corollary of this is that the rural workforce will decrease. The growing labor deficit would prompt farmers to adopt automation increasingly, whatever their conservative attitudes might be.

Robotics, remote, and AI technologies can perform many tasks traditionally performed by people. Intelligent software platforms can already monitor and control multiple remote facilities. Automating operations can save time and improve performance by avoiding human error. Many developed world farmers already use various innovative technologies and cope with the challenges of complex software systems. In turn, many software systems can process and understand human verbal commands, eliminating the need for farmers to deal with complicated graphic interfaces.

Satellites and drones also play a role in the agricultural revolution. Data from these promises to detect worrying soil and plant conditions and spot crop diseases faster and more reliably. Software for rapid imaging of crops has been available for decades. More research remains to be done before AI and machine learning systems can make full use of it.

Agriculture is subject to greater government and public attention and interference than most other industries. Satellite data will also play a role in this. Thus, it can be used to monitor indicators that underlie targeted government and private funding decisions. In extreme weather and natural disasters, not all farmers will experience equal difficulty, nor marshal equal resources. Satellite imaging can provide meaningful leads that can help distribute aid or subsidies more equitably and purposefully.

Despite their promise, innovative and AI technologies still face many challenges in agriculture. Extensive research is required to create feasible universal IoT or AI platforms suited to every farm and every farming culture. Agriculture is a minefield for statistical quantification, even if performed by leading researchers. No two fields and no two sets of conditions are the same.

Moreover, conditions are constantly evolving, even across a single field. At the current state of the art, a machine learning platform intended for universal farming use would need an enormously complex algorithm. It's difficult to imagine such an algorithm will be developed anytime soon. Ultimate commercial return will be the real driver in the research effort.

IBM foresees a market for over 75 million network connected farming IoT devices like sensors, mobile robots, and agricultural drones. IBM researchers see an average farm generating some 4.1 million data points daily by 2050, from only 190,000 in 2014.

Used smartly, this vast amount of data and the huge army of robots and computers that will gather and process it can boost agricultural yields and food quality, cut energy consumption, and spread available resources much more equitably.

MANUFACTURING

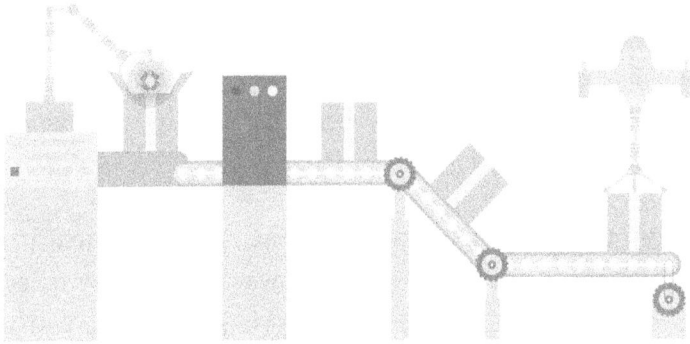

General Motors put Unimate to work in the early Sixties. The world's first industrial robot had a fitting name for something unique and quite 'matey' to its human 'colleagues.' It did not stay unique for long. Initially fit for only simple tasks, robots quickly moved onto more complicated ones. They also began to replace their human colleagues. So much for the 'mate' in Unimate!

The next stage of manufacturing automation will harness machine learning. This will place Unimate's progeny at the top of the manufacturing pyramid. The 'usual suspects' to watch in these emerging technologies include sensors, computer analysis, and cloud-based technologies. A feature of theirs will be the ability to take autonomous decisions and even repair themselves by diagnosing and replacing worn or defective parts. Systems endowed with AI would be able to keep minute track of items and procedures to an extent humans have always found (and will always find) vexatious.

This, in turn, would reduce error. Human error costs in manufacturing are high and can cause risks and reputation damage. Extreme cases have included aviation and even space exploration disasters and embarrassments. Alongside preventing human error, AI systems would diagnose completed products and prompt, timely preventive measures.

Marrying sensors, monitoring systems, and manufacturing equipment is not an easy task, however. A factory using AI systems would be a complex tangle of innovative technologies sharing a common network. Such a factory would integrate equipment of great diversity even on a single production line. Designing such an intelligent system would involve much mental effort and planning. Even a single sensor would be critical in an AI manufacturing facility. If one lacks the right data, even for a very simple process or task, one would be unable to take the right decisions about the overall process.

The need for flexibility and redundancy would add to the above complexity. Equipment and systems would need to be replaced independently of one another as they age or break down. Moreover, such replacement would have to be 'seamless,' rather than disruptive.

Similar challenges will be familiar to those who worked on computer networking a couple of decades ago. They were, of course, overcome to the extent that most of us take today's World Wide Web for granted. The likely demand for smart manufacturing alone should resolve all of the above challenges. A *TrendForce* report foresees a market worth $250 billion in 2018 in the field.

Big Data and AI will be the drivers behind the next industrial revolution in which AI-powered systems will replace today's simple task manufacturing tools. But software and the underlying algorithms will remain crucial for any manufacturing AI system. Today's software and algorithms are designed for set purposes. Assuming car manufacturing software can be reprogrammed to manufacture aircraft, its algorithms would still need to change to keep track of different tasks and hence tackle decisions unique to aircraft making.

Take for instance designers creating 3D-printed part prototypes and checking how well they work with other components. They currently need software attuned to their industry. In the future, however, they would be able to make use of AI's ability to learn. AI platforms would be used as universal tools to check or test products and systems across industrial demarcations.

Whatever testing went on before, the most important part of any manufacturing process is production itself. Here, one needs intelligent robots that can control any task on the production line and can rectify any issue with defective parts or incorrect operation. The technology for such intelligent systems is already becoming available and has been implemented by many manufacturers.

The problem is that no single AI platform is currently able to perform universal tasks related to any manufacturing process. Hence, each manufacturer must develop their AI-powered systems or tweak available ones to suit their needs. Reliable state-of-the-art AI systems can spot defective components, order new ones, and replace them automatically. Yet, a truly intelligent system would have the ability to make much more complex decisions that would even encroach on human managerial authority.

But AI technologies and intelligent robots will also lay workers off. This elevates smart manufacturing to the social and political scene. It may be argued that manufacturing AI platforms draw significant research and development investment which creates jobs. These jobs, however, will be white collar jobs. Their number would be significantly smaller than that of the blue-collar jobs lost. The number of people qualified for them would be significantly smaller than that of people qualified for the blue-collar end of the spectrum.

Finding the right balance can only be a matter of time. Only experience will show how well manufacturing genuinely copes with new technology at its elbow and how well society genuinely copes with the whole concept of robots replacing people. The great dilemma is how to integrate and retrain manufacturing workers whose only skills are easily replaceable by intelligent robots.

Today's technologies allow manufacturers to automate production, packaging, and even clerical tasks. But white collar workers should not be complacent! Scientists are growing increasingly certain that AI-powered systems will begin encroaching even on areas remote from manufacturing. Today's brand designers and planners, product architects, process schedulers and controllers, production supervisors, monitoring and evaluation specialists, component integrators, quality controllers, stock managers, and batch controllers, among others, all stand to see their working – and leisure – lives changed.

The skill-based economy is knocking at the door. Over the next decade or two, we shall witness huge changes. New trades will emerge, swathes of people will lose their jobs, and skilled workers in all fields, including manufacturing, will become more attractive to employers.

Two things are certain. Regardless of political and social movements opposing progress, like the Luddites, historically they have proved irresistible. They have also proved painful.

RETAIL

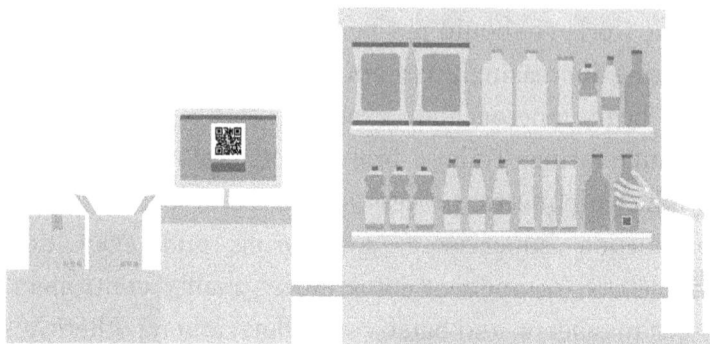

Retailing's classical incarnation included a street, a shop, a salesperson, items on sale, and customers. Automation and electronics first came to the scene some 50 years ago with electronic tills. Spreadsheets followed. E-commerce came in the 1990s. The first static merchandise websites began forcing traditional shops out of business by the first decade of the 21st Century. Chatbots and predictive analytics began replacing salespeople by the mid-teens. Interactive online shopping is far from the end of the road, however.

The online retail boom of the past two decades began to transform retail by borrowing technologies from warehouse management. In them, software typically tracked stock levels and individual purchases, ultimately providing data. Managers then used this data to order and replenish stock. More sophisticated software could also use networked platforms to monitor and manage more than one warehousing or retail location.

Now AI and machine learning are about to invade retail. They are already available. Retailers simply must take greater advantage of them. Retailers are very aware of the industry's coming technological revolution. A Blue Yonder survey shows 77 percent of UK grocery retail directors acknowledging that AI and machine learning will usher wholesale change; 23 percent state they have already invested in these technologies.

AI's essential promise is that it can automate decision making. Rather than a human decision maker wading through raw and often confusing data while perhaps burdened by existential troubles of their own, AI promises to process data as soon as it arrives and to take instant decisions that would be entirely sound.

In retail, such automated decision making would optimize strategic areas like pricing and replenishment. Meanwhile, machine learning would align decision making with key performance indicators like margins, volumes, and markdowns.

Most retailers already use software with some AI capabilities. This is mostly in the crucial discipline of out-of-stock management. Customers who fail to find an item at one retailer are unhappy and tend to go to competing retailers. On the other hand, retailers who have too many items freeze cash that they could have spent to better effect elsewhere. By honing the old art of just-in-time, AI would help avoid both overstock and understock scenarios. Machine learning would allow retail platforms to analyze historical data. They would then plan and marshal stock levels down to individual locations and periods.

A problem many retailers face is that although they might have invested in a stock item, it might be stuck in a backroom and away from the eyes of browsers. AI retail platforms would help retailers boost sales items on shelf availability. A recent Quri study found that some major summer brands in the US were only available on shelves for as little as 43 percent of the time. This self-evidently lowers return on investment for retailers markedly. It also irritates their customers, which is worse because irritated customers play into the hands of competing retailers. The intelligent retail software of the future would spot on-shelf availability gaps easily and generate early warnings. Moreover, given the right marriage to mechanical warehousing equipment, it would even get them filled without human intervention.

Predictive analytics is already part of all good retail software platforms. The game changer, however, is the integration of AI-powered software modules with machine learning capabilities. Such a platform would allow retailers to understand market trends, optimize inventories, and even run putative scenarios to underpin intelligent and justified decisions in advance. Thus, a traditional brick and mortar shop could simulate diverse weather and weekday scenarios to glean likely future sales from historical data. Its managers could then adjust inventory and staffing to cut costs and boost sales.

'Heat maps' showing what elicits the most clicks from online visitors are already common in e-commerce websites. Coming technology will allow AI-powered systems to track more subtle brick and mortar store visitors behavior, showing where shoppers tend to linger and where they tend to make impulse buys. A retailer could analyze traffic patterns, tweaking shelf layout and product position accordingly. They can also analyze shoppers by age, sex, the number of items purchased, and purchase value, which could suggest multiple improvements in product positioning and shelf location.

Another application of machine learning and AI would employ beacons and geolocation. Retailers would be able to send messages to mobile devices of nearby browsers, while an AI-powered application could even beam customized messages trimmed to private preferences and buying habits of individuals. It goes without saying that messaging policies ought to be weighted carefully; certain customers would be irritated by it. Ethically speaking, customized messages on discounts or special offers should be deemed entirely acceptable if they reached people who had willingly signaled themselves as potential buyers by entering a shop and browsing its shelves.

Implemented piecemeal, the above technology would benefit retail little. Its real weight comes to bear when it works jointly in a complex retail platform. Such platforms would process and analyze data from points of sale, e-commerce sites, mobile platforms, customer relationship management systems (CRM), and social networks. They would then use AI engines to generate and send the right sales messages to the right persons at the right time, through the right channels.

Retail is a data-rich industry. It collects piles of information every day. Only AI and machine learning can make timely use of this information. The human commercial hunch will remain unmatched, but it's too prone to subjectivity and is too slow. AI retail systems would process and analyze hundreds of factors and indicators within seconds. They would then take decisions with a rapidity that would challenge any person. They would replenish stocks and complete online purchases while also optimizing the e-commerce site to personalize each customer's experience.

Chatbots have been available for some time now. They understand and speak a natural language and can utter spoken prompts based on customers' answers or preferences. Impressive advances in AI image recognition opens further possibilities. Customers would be able to post a product's image into a search engine and then find online retailers who offer that precise product. Once they have located an online store, an AI method known as 'online learning' would allow the software to track and analyze each click in real time to personalize their experience based on strictly individual preferences and interests. This copies the behavior of customers in a traditional shop. There, they scan the shelves for products, pick up and examine goods, and may consult salespeople on the features of specific goods. The difference is that now they would to do this in their homes and that the shop assistants would be virtual bots powered by AI and machine learning.

Evolutionary algorithms (EAs), a class of AI methods, promise to take retail a step yet further. They can adjust website elements like typeface, type size, images, page sequencing, messaging options, and even page layouts to suit different audiences and even individual visitors. EAs would evaluate the performance of website variations and generate new ones to boost sales. This mix of Big Data, machine learning, and AI-powered predictive analytics would open the gates for a new age in retail where every single online store visit would look subtly different.

Instant adaptive personalization is not only the future of online retail. Much of the technology behind it can happily go into traditional shops. There, it would navigate customers straight to the products they are seeking through their mobiles or by beaming specific audible recommendations at them. This omnichannel retailing, where retailers can interact with customers both online and offline, could result in up to a 30 percent higher lifetime customer value. It would also dovetail with stated preferences of today's customers, which are for personalized experience online, by mobile, and in store.

AI and machine learning can deliver this personalized and highly customized retail experience. The trend is already in place and is set to become irreversible. A report by

Gartner says that by 2020 customers will conduct 85 percent of their commercial interactions, including retail ones, without contacting another person. Another survey, by Statista, reveals that only 22 percent of US household shoppers aged 18 to 29 prefer to shop in stores. The other 78 percent prefer online and mobile shopping. This indicates the future trend all too clearly.

Google, IBM, Yahoo, Intel, Apple, and Salesforce are in fierce competition to acquire AI and machine learning startups. Many such startups are developing solutions that can be implemented in retail. The first quarter of 2017 alone saw more than 30 of these acquisitions. All major AI players are investing heavily in advanced AI technologies like deep learning, neural networks, and cognitive computing.

Machine learning and Artificial Intelligence technology are finding applications across a swathe of industries. Though retail lags somewhat, the next decade stands to witness a rapid hike in AI techniques. They will have Big Data, predictive analytics, and adaptive retail systems provide customers with completely personalized shopping experiences. The first big step will be for stores to employ natural language and AI to adapt to each customer using machine learning.

FINANCE

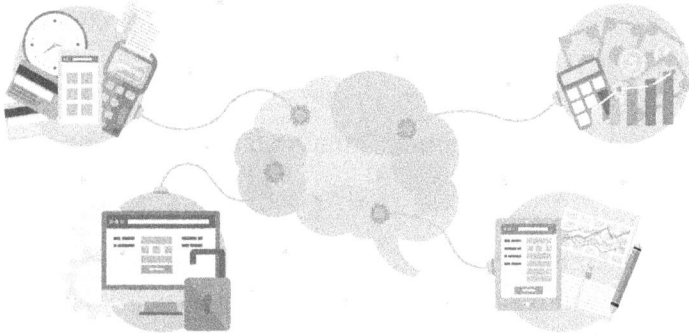

FinTech startups are discomfiting traditional banks and financial institutions. Innovators after a slice of a lucrative market are changing it by disrupting its venerable practices. In 2017, FinTech market transaction value should reach $3.3 trillion. An impressive 20.5 percent annual growth by 2021 should push this to $6.9 trillion.

For centuries, finance houses have profited from lending and payments. But today, digital lending and payments are bypassing tellers and bank branches. That new wave of Eighties banking, ATMs, looks antiquated as AI conquers the field.

FinTech's largest segment is digital payments. Statista estimates its worth at $2.6 trillion in 2017. They forecast that 3.5 billion customers will use digital payment solutions in 2021. Many of these clients are the prime customer segment of legacy banks: 'corporates.' Not blind to cheaper and faster online payments, by opting for them they leave the embrace of banks.

AI and machine learning algorithms power much of FinTech. They can adapt to needs and behavior patterns of customers. To customers, this means personalized online service that appeals especially to the younger generation. Banks ought to view this with alarm: older clients who prefer human interaction cannot be around forever.

So much for the base level we still know as retail banking. Change is also afoot in the upper hierarchy. By the mid-Nineties, banks and investment houses started pouring money into AI systems which could take investment decisions. However futuristic it seemed at the time, this effort resulted in today's forex, stock, and commodity trading systems. AI tracks and analyzes market trends and generates buy or sell prompts. Some systems, particularly in forex, are even allowed to make small or not so small trades on their own.

The trading bot revolution has become so comprehensive that many a lay individual has opted to make a buck or lose their shirt while trading in the comfort of their own home. Admittedly, their example is extreme. Lone traders can only rely on beginner's luck when competing with hoary banks and funds. The trend is telling: each home trader chips away at the livelihoods of exchange brokers. Today's trading bots are rather crude, however. AI's promise is much greater.

To grasp that promise, it helps to understand AI better. Big Data is a term we shall encounter more often. Finance, like most other industries, has always had access to huge amounts of sheer data. At its retail end, everything from the effect of soccer championships or the weather to petty customer behavior traits has been available for the taking. Management attention has focused on targets and limits, harvests, manufacturing and oil outputs, all the way to intricate indicators emitted by domestic, regional, and world politics. Up to now, most of this data has at best entered the haphazard occupational lore of bankers. It has lain unprocessed because its sheer weight rendered it unprocessable.

Processing and analyzing this data is at the heart of AI's promise. But where AI really wins is in what follows from there. Having crunched the data, AI can suggest better decisions. A minor step further takes it into territory where it may act autonomously. Perhaps it might jump at a huge split-second opportunity the like of which was often impossible to grasp in the past. Or, moving from the sublime to the trite, it might bend institutional rules to soothe a livid customer.

Even now, mobile apps can process Big Data and provide insights previously guarded within the closed guild of investment bankers and brokers. Moreover, these software apps are adaptable. One can set targets, limits, and triggers and manage them in accordance with advanced algorithms.

More and more FinTech startups are rolling out surprisingly sophisticated services. Instant payments, crowdfunding, invoicing and accounting, and complex cross border transactions are all there for the asking, as are investment 'robo-advisors' and wealth managers. Some solutions even mimic the capabilities of enterprise software platforms and are free to boot.

Conservative finance niches like insurance are feeling the pressure of insurance underwriting AI systems able to suggest better decisions through crunching vast amounts of data. Within a decade or less, no crucial financial decision will fail to feature AI, machine learning, and cognitive computing input.

Furthermore, most e-commerce and payments solutions already employ AI to detect fraud. This minimizes financial transaction risks. Some finance houses use AI to run capital planning scenarios and flag transactions for compliance review.

In the next few years, financial, insurance, and regulatory technology (inevitably termed FinTech, InsurTech, and RegTech) will move into segments like asset and wealth management, capital markets, digital cash, and treasury functions. The technology is gaining ground so fast that legacy institutions are challenged to keep up with it, finds a PwC report. Most banks are trailing behind in providing the completely personalized experience a growing number of their clients demand. Finance desperately needs to keep customers on their side which calls for outside-in, as opposed to inside-out, insight: something only AI can provide.

The old business model is unraveling. Financial innovators are poised to feature much larger in the new one. They already use AI. They are already abreast of customer demands and behavior. They already anticipate even minor changes and trends. They are at home in the increasingly mobile infrastructure of communication channels, call centers, chat lines, social media profiles, and mobile apps. Yes, privacy and data security concerns inherent in that novel environment trouble them much more than they trouble legacy banks. It is the coming generation's preferred medium for conducting all financial transactions. Now, banks might appear large and intractable but can be surprisingly agile. Many are rolling out online services. Some are using voice recognition as a way of making transactions unimpeachably secure. Not a few are cutting costs by applying AI-powered automation. They understand Millennials are accustomed to virtual assistants and mobiles.

Ultimately, however, techno wonder must be put into context. Current AI technologies have real limitations. The vexed area of security and data protection is a very prominent example. At the workaday level, more complex queries and transactions can quickly hit the figurative stops. Qualified human assistance is still welcome, quite apart from the reassurance it brings, not least to older customers. Banks and finance houses are thus seeking the elusive and shifting point of proper balance between automation and human intervention. Finance is unlikely to become as robotized as farming or manufacturing. At best, robotics can complement and support decision making, cut interaction cost, and boost overall efficiency.

Predictive analytics and machine learning is the likeliest area of AI expansion into finance. It has multiple applications ranging from portfolio management, through algorithmic trading and fraud detection, to lending and insurance underwriting. At the retail customer level, AI's improving natural language capability will enable more and more on-the-fly transactions and cut personal visits. AI will also analyze natural language news, distill trends from it and highlight likely market sentiment. This will extend its current role of deriving trends from market movements.

What lies beyond AI in finance? The clever money is on cognitive computing and blockchain technology. Emerging expertise there would one day allow users to employ self-programming algorithms. These are the technologies that, in their turn, will obsolete today's disruptors.

ENERGY

The energy and energy distribution industry is a field where artificial intelligence has yet to be deployed on a large scale. But even at this stage, all signs point to AI and machine learning technologies offering unparalleled benefits.

The use of AI, machine learning, cognitive computing, and neural networks across the energy industry is inevitable. It must be, seeing as grids are becoming unmanageably sophisticated. Another factor that will spur the adoption of AI is the focus on new energy sources (think renewables) as well as the rapid transformation of the traditional supply/demand model of energy distribution and consumption.

On the customer service side, a good number of energy companies already employ chatbots. These relatively simple AI systems can handle basic customer questions, using machine learning to improve answers over time. In this way, an energy company can divert resources from customer service operations to its core business, which typically involves developing sources or distributing energy.

Obviously, this is just a starting point for the application of advanced computer technologies across the energy industry. At stake are far more important issues such as the optimization and automation of grid maintenance and drilling operations.

A series of regulatory measures have been adopted around the world. As a result, many countries have made it mandatory for renewables to account for a certain percentage of power generation. Renewable energy resources, however, largely depend on weather conditions:

you need sunshine and wind, for example. This forces the conclusion that using renewables efficiently hinges not only on energy demand estimates but also on accurate weather forecasting.

The use of machine learning algorithms where AI processes and analyzes weather data can improve the efficiency of a power-generating unit that exploits renewables. Such an AI-powered system will also analyze historical consumption data to match estimated demand with projected output and optimize the planned demand and supply.

Then we have a growing number of municipalities, businesses, and even individuals becoming small-scale energy producers. This makes the grid more sophisticated in terms of energy distribution control and supply/demand management. It's beyond our abilities to predict correctly when a household with installed energy sources will switch from consumer to supplier and back again. A specifically-designed AI system that features machine learning capabilities can track consumption and supply over long periods, and then combine data on overall supply and demand to predict such events accurately.

Predictive analytics should be an integral part of such systems if we are to achieve maximum efficiency. It is needed because this type of consumers-producers has varying and largely random patterns of individual production and consumption. Consequently, a future energy AI system should also manage thousands, if not millions, of producing-consuming end points by gradually stocking up knowledge about their energy usage and production patterns. At a national grid level, the AI system must ensure a supply/demand balance at all times and without faults. The next step will be the creation of self-healing networks with zero fault tolerance. They will be able to manage millions of small consumers-producers, be it within national or global energy grids.

Let's consider DeepMind as an example. This AI algorithm became the property of Google through the purchase of UK start-up DeepMind Technologies, Ltd. DeepMind could be used to optimize and manage the UK national electricity grid, according to media reports. Such a bold step is driven by the emergence of new, disruptive business models in the field of energy sourcing and distribution. Peer-to-peer energy and community energy are examples of such business models. Although still in their early stages, they will require the adoption of advanced technologies only AI-powered software can provide. Last year, for instance, DeepMind managed to reduce the overall power consumption at Google's data centers by 15 percent. Digitizing the entire energy system is inevitable given the complexity of emerging energy grids and the growing importance of Internet of Things (IoT) networks, where billions of consumer and producer devices will be connected to the grid and other devices.

However, automation and digitization are not limited to energy distribution and electricity grids management. Advancements in robotics, 3D visualization, sensor technology, and graphics processing can bring the end of drilling and oil pumping as we know it. Many oil and gas companies, large and small alike, already implement AI technologies to ensure accurate and optimal drilling. The use of AI holds potential for increasing production without workforce expansion. This, in turn, lowers costs and enhances accuracy indicators by eliminating human error.

Oil giants like BP are developing AI solutions internally to combine datasets that cover sensor information about flow rates, pressures, equipment vibration, and other metrics with data from sensors that are monitoring the natural environment. By combining data from equipment with data about seismic activity or ocean wave height, they can optimize the entire process of organizing and performing onshore and offshore drilling operations. So, a well-designed AI system equipped with proper sensors and procedures can improve efficiency and safety in drilling. Engineers at oil and gas companies are also looking at cognitive computing as a technology to embed into such systems for maximum effect.

AI systems in oil and gas drilling and pumping can also perform reliable sensor diagnostics and control parameters for temperature, pressure, and phase composition. These indicators are constantly monitored, but predictive analytics, combined with self-learning and self-healing systems will disrupt the entire process of drilling and pumping. Eventually, these systems will evolve into AI control clusters able to make complex control decisions based on flexible, non-deterministic strategies. Also, the use of full-fledged AI systems for monitoring and managing drilling operations will spawn a generation of alternative designs that integrate existing process configurations and introduce optimized solutions.

By and large, AI will play an increasingly important role in optimizing and managing both drilling and power generation and energy distribution. The management of future energy producers-consumers, or prosumers, will require sophisticated AI systems where IoT technologies are applied to control local power sources and local networks connected to a national grid. Energy storage management, a technology gaining pace globally, also requires advanced software to ensure the entire energy system is reliable. Self-healing grids are also inevitable since current automated systems cannot cope with the growing complexity of energy generation, distribution, and storage. Energy load balancing is another crucial field where AI will help achieve sustainable development.

The integration of smart devices within the power generating and energy distributing networks will deliver a number of benefits. These include markedly greater efficiency, proper fault and emergency management, as well as a personalized two-way energy flow. The emergence of this "Energy Cloud" will boost AI adoption across all segments of the energy industry. It seems safe to assume that machine learning and predictive technologies fused into an AI platform will become mandatory for future power generation and energy distribution networks.

HEALTHCARE

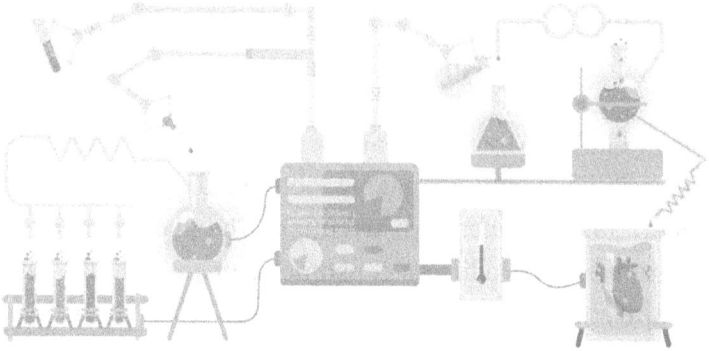

Healthcare is something each of us needs. Demographics will ensure it will not only continue to be a hugely politically and economically sensitive area but will become ever more so. Healthcare establishments have used IT applications like remote patient monitoring, managing medical records, and organizing patient routes and treatment plans for some decades now. AI-powered equipment, however, is still rare in the field. It remains expensive and is yet to undergo the exhaustive multiple proving cycles required in the field. But this is starting to change.

Disruptive technologies like machine learning and cognitive computing are spearheading broader AI ingress into healthcare by promising to reduce the cost of high tech medical applications.

Over fifty startups have entered the healthcare AI systems market since January 2015. A CB Insights report shows more than a hundred companies in the field. They are applying machine learning and predictive analytics to develop drugs faster, diagnose diseases through automated medical image processing, and provide remote monitoring or virtual assistance to patients.

AI is likely to have a greater impact in several healthcare areas in the foreseeable future. Some of the technologies and applications are already available, though most need further honing before they can become mainstream and gain the confidence of patients. AI systems can crunch Big Data in storing and processing virtually any amount of patient records and treatment plans. Current software can recognize natural language inputs and commands reliably: useful for entering a clinician's notes and data into any medical system. Doctors can retrieve all medical data on patients with a simple voice command which is beneficial in monitoring patients remotely or treating them online. Some companies have already deployed these technologies, and this segment promises to grow fastest over the next few years.

Data management is one of the most powerful applications of AI. It uses data lakes to gather and store medical data. Google's DeepMind Health project, for example, is aimed at mining medical records data, while IBM's Watson is providing clinicians with evidence-based treatment options. Cognitive computing, though in its early stages, can equip doctors with analytical and reasoning capabilities provided by virtual assistants armed with a broad range of clinical knowledge. AI tools able to spot even tiny deviations in radiology images are gaining ground, for example.

Significant human intervention is still required in all the above applications of AI. But one can imagine a future medical platform able to perform many medical tasks with minimal or no supervision. Machine learning is the technology that will make this possible. An AI healthcare system can process myriads of clinical records almost instantly and go on, under supervision by scientists and learning from historical data and mistakes, to produce sound algorithmic analyses of most diseases.

Researchers already know how to program a learning algorithm. The bigger problem is getting patients to trust a machine for their treatment. Once this prejudice is defeated, deep learning performed by a program designed to analyze every part of the body can deal with a broad range of diseases, experts believe.

The perennial short supply of medical professionals is also prompting healthcare to look for relief in AI. A World Health Organization report forecasts a shortage of about 13 million healthcare workers by 2035. A possible and obvious solution is to adopt AI-powered systems and healthcare virtual assistants that can close that gap. Even existing technology can affect simple tasks like assisting the average patient in taking their medications, providing basic medical advice, and monitoring one's health.

Another healthcare benefit AI promises is to enhance drug development. The traditional research and development cycle of a new drug through clinical trials sometimes take decades. The cost of drug development is often massive, putting the cost of new medicines beyond the reach of many. Now, since medical AI systems would play with vast amounts of data, they could direct some of it into pharmaceuticals development. Employing AI there would not only shorten the development cycle for new drugs but also make them much more widely affordable.

AI healthcare assistants might also be able to perform early diagnostics, a cornerstone of any medical treatment. For example, the Watson platform gave the same recommendations as expert oncologists in 990 of 1000 cancer cases during testing at the University of North Carolina School of Medicine.

This is just an example of an AI healthcare platform with some deep learning capabilities. Known as black box algorithms, these AI systems can come up with diagnoses, decisions, and forecast outcomes without anyone having to micromanage their algorithms. Though this ability is initially slightly alarming for patients, scientists cite the example of aspirin. The therapeutic mechanism of this most widely used medication of all time remained unknown for nearly a century. AI systems that feature deep learning capabilities find and create their own rules and are already able to outperform human doctors in diagnosing certain complex diseases.

These innovative technologies lay the foundation for the extensive use of AI, machine learning, and cognitive computing in literally every healthcare sphere. Nonetheless, there are concerns and uncertainties on how AI would shape the future of the healthcare industry.

The biggest question is one of the ethics. Will AI tools evolve from their currently foreseeable role of suggestion providers to become decision makers? The question, common to all AI applications, acquires much greater sensitivity when it comes to human health and lives. AI accountability is the main concern. The issue might be resolved through wider application of deep learning and the gradual improvement of machines' decision-making capabilities.

Security and privacy risks are another concern. Patient data is extremely sensitive, and patients' records require even greater protection levels than other personal data. Privacy pertains to the broader domain of protecting any sensitive data in an interconnected world. But data manageability in healthcare must obey much stricter rules that those obtaining elsewhere. The issue comes down to strict policies strictly enforced.

Resolving the biggest problem above – that of decision making by AI systems – could simply be a matter of time. It would take time for people to outlive their intrinsic prejudice towards mindful decision making by inanimate objects and machines. This would probably occupy the lifespans of several generations. AI systems' overall communicating and interacting capabilities would improve beyond recognition in that time. Programming a machine to demonstrate empathy, required in healthcare communications between a patient and a doctor, is easy.

The difficult part is to create an algorithm and technology that can convince the human patient that this empathy is real.

ART & MUSIC

When we think about AI, we often refer to those areas and industries where humans may be easily replaced: manual labor, driving, medical diagnosis, or business analytics. However, there is now a new wave of interest in the application of AI in creative industries, such as art and music–spheres that, until very recently, were considered to be exclusive domains of human creativity. Experiments with AI in musical composition and art have been done since the 1970s; however, recent breakthroughs in deep learning paved the way for more sophisticated AI systems that have a potential to disrupt contemporary art and music industries. In deep learning, complex artificial neural networks (ANN) organized into multiple layers extract hidden features and patterns from musical compositions, sheet music, and works of art. Algorithms used in these networks allow computers to find their way through data and come up with non-trivial and unobvious solutions. As a result, listeners and art critics are witnessing something that may be described as Artificial Creativity.

So, what does it take to be a good classical music composer? Many people would say that talent, musicality, and creativity are essential prerequisites for this. Along with intelligence, these are integral features that make human beings unique. Therefore, it may be disturbing for many to witness the emergence of such AI as AIVA (Artificial Intelligence Virtual Artist). AIVA can create classical music that sounds as if it was written by a talented human composer. In fact, AIVA is the first AI ever to acquire an official status of a composer. What takes a lifetime for a human composer to accomplish, AIVA learned in several months reading through a large database of classical works written by Bach, Beethoven, and Mozart. During the training, AIVA learned various musical styles, compositional techniques, instruments, and ways to put them together in musical tracks, which are now accessible via popular streaming platforms, such as SoundCloud.

Musical AI such as AIVA may open an era of automatic music composition and production, but their impact goes beyond that. Contemporary AI can create new musical instruments and sounds never heard before. For example, Google-backed NSynth (Neural Audio Synthesis) project uses neural networks to learn mathematical 'vectors' of various instruments, which are then synthesized to produce entirely new instruments and sounds. The AI can merge hundreds of musical instruments into one, and create countless new sounds from those we already have. Although blending musical instruments is nothing new, NSynth pushes this old practice to a qualitatively new level. Composers, individual musicians, orchestras, and the audience will benefit from this unmatched musical diversity delivered by musical AI.

Also, musical AI is becoming a useful tool for learning and performing music. One example of this trend is Popgun, a startup that is developing the first AI musician who learns from human musicians, plays music with them, and makes his creative contributions into original musical pieces. AI developed by Popgun can help arrange compositions, or suggest instruments that best convey musical ideas and melodies composed by humans. Another self-learning system, Pacemaker, is advertised as an AI DJ that can create digital mixes and remixes from files and streams. Harnessing the potential of these AI systems will change the way people produce commercial music these days.

Musical AI is driving huge changes not only in the way music is produced but also how musical businesses monetize and distribute content. Growing popularity of online streaming platforms, such as Spotify, SoundCloud, or Google Play Music, causes gradual replacement of the outdated model of the physical distribution of music. With loads of music available in online databases and streams, listeners are struggling to navigate through the infinite world of the digital culture. AI techniques used in music recommendation systems aim to solve this problem.

AI-enabled recommender systems, like Musical Genome Project used in Pandora automated music recommendation service, try to predict musical preferences and tastes of users by studying similarities in tracks they listen. Each musical piece in Pandora has 500+ features, such as harmony, melody, or voice, and AI tries to identify the ones which are the most frequently met in tracks preferred by users. Another musical AI named Robin can simplify the life of music lovers by reserving and securing tickets to live concerts. Also, it helps music companies become more competitive by collecting and analyzing real-time demand data for live events and albums, forecasting prices and predicting the success of new musicians, albums, and live events.

Musical AI may also have a new role in the organization of music business itself. To attract and retain musicians, companies, and audience, online music streaming platforms urgently need new business models for royalty transfer, licensing, and monetization. AI-enabled blockchain technology (e.g., Bitcoin or Ethereum) may be an excellent solution to this problem.

Blockchain technology allows creating transparent, safe, and flexible transaction systems that are suitable for processing user micropayments for tracks, and supporting royalty transfer and licensing payments online. For example, AI created by JAAK startup analyzes metadata on song ownership and licensing to disburse money to owners in a more efficient and decentralized way.

In addition to content and transaction management, state-of-the-art machine learning algorithms may be used to analyze user behavior in audio content networks to produce actionable insights for music companies needed to improve online music production and distribution.

This data may be used to enhance monetization of musical content by inserting personalized ads into audio content and podcasts. Such technology is already offered by Pippa, a company that uses AI to perform deep audio search and ads personalization based on the content of podcasts.

Harold Cohen, a pioneer of computer-generated art in the 1970s, believed that it would be hard to create a machine that can make something approaching art in what is left of this century. Scientists, however, continue to push the boundaries of AI creativity. As an example, Google, Inc. recently showcased its Deep Dream AI that uses deep learning to create its art. The program is trained to identify objects by scanning millions of photos. It starts from low-level features of images, such as points, angles, shades, and ridges, and moves to more abstract ones, such as objects and landscapes. The program can recreate random composites of objects to produce new images.

Researchers can manipulate Deep Dream's creativity by activating certain neurons. For instance, if they excite the neural net's 'cat' neuron, AI would search for cat features in the landscape and maximize them. According to Steven Hansen from Google's Deep Mind, "It's almost like the neural net is hallucinating." To the surprise of critics, AI-created art becomes popular among art buyers; twenty-nine paintings created by Deep Dream were sold at a charity auction in San Francisco with the priciest artwork receiving an $8,000 winning bid.

Despite doubts over the ability of autonomous AI to revolutionize the art market, it is now evident that art AI may become a useful tool for art practitioners and learners. According to Francesca Rossi, a researcher at the IBM J.T. Watson Research Center in New York, AI, "can really help humans enhance their creative capabilities, because even our creative capabilities don't come out of the spur of the moment, they come out of analyzing the world around us, a lot of data around us."

For example, AI can automate the routine part of artists' work. Back in the 1970s, Harold Cohen used *Aaron* AI to colorize his abstract paintings. In his words, the program could produce 50-60 excellent images every night when he left it running. In the same vein, the AI team at The Carnegie Mellon have recently created a bot that works as an invisible coloring book. The robot has a paintbrush attached to its arm that guides the human hand to meet the instructions of the AI computer program. The robot helps an artist to steady his/her arm by pushing it back in the necessary direction, if an artist moves away from the curve specified by the computer. In this way, AI becomes an artist's assistant.

AI also excels in what computer scientists label style transfer – transferring style from one visual artifact to another. This technology is now widely used to imitate styles of paintings and genres (e.g., horror) in video games and animation. One interesting project that probes into this direction is MIT Nightmare Machine that uses deep learning to transform famous buildings, such as Big Ben, the United States Capitol, or Tower of London, into haunted locations like those one may see in horror movies. To create horror imagery, Nightmare Machine trains on thousands of pictures retrieved from horror movies. Over time, AI learns horror aesthetics and develops strong neural connections that may be easily excited to create haunting images from any content.

Style transfer technology used in mobile phones makes art creation accessible to a wider audience. Such mobile applications as Prisma, Comet, Pikazo, and BitCam, can transform user images into Impressionist or Cubist paintings expanding the way social media users express themselves. In turn, AI-based analysis of these user-generated images allows companies to gain a deeper insight into what visual elements in their marketing campaigns are the most exciting and engaging for consumers. Marrying art AI and business analytics, companies will have an opportunity to find out what images sell more products, and what colors, and types of visual content best capture users' attention.

As these examples vividly demonstrate, contrary to what critics say, state-of-the-art musical and art AI, can become a powerful source of human creativity and driver of creative industries. AI capable of analyzing art and music can give people a better understanding of our artistic creativity that is often hidden under the hood of vague concepts of intuition and artistic freedom. At the same time, however, AI capable of producing decent art and music will raise our standards of good art and make it more challenging for people to create valuable things.

MEDIA

Over the past two decades, media industry had to respond to numerous technology challenges. First, the emergence of Internet and digital content delivery jeopardized print press, TV, and radio. Media industry adapted to these changes by introducing new IPTV, OTT (over-the-top content), and digital media models. Then, social networks heralded the Web 2.0 era of interactive content, and deeper audience engagement in media production and distribution.

Mirroring this trend, media had to become more open, social, and flexible. AI is the next disruptor of the media industry that brings both new opportunities and challenges. AI in the media industry is fueled by the proliferation of big data, data mining, and machine learning algorithms that revolutionize techniques of content generation, delivery, brand management, marketing, and business analytics.

Media companies are seeking new ways to use data-rich digital environment, cloud infrastructure, and machine learning to offer highly customized, flexible, and relevant content to their consumers. The impact of emerging AI technologies is especially strong in online news and magazines, TV broadcasting, movie industry, and social media.

Journalists and writers are gradually losing their monopoly over the written content. Natural language generation is a subfield of machine learning that stands behind this revolutionary change. Modern algorithms teach machines to construct descriptive and explanatory texts that make sense to readers. AI that can express itself is already used by the leading online newspapers and magazines. For example, since 2016 the Washington Post has been experimenting with automatic storytelling using its in-house Heliograph AI. This software was employed in the coverage of Rio Olympics and the US Presidential election in 2016. Heliograph constructs news reports by putting relevant data and facts into language templates. Powered by data mining techniques and ML algorithms the software can convert statistical data, diagrams, graphs, stock market prices, and other data-rich content into descriptive reports.

Automated storytellers have a promise of automating many tasks usually performed by news reporters and journalists. However, people behind these innovations believe that such AI as Heliograph, in fact, could free up reporters and editors to add analysis and real insights to stories rather than spending countless hours publishing results or news.

AI in online media may be used in other innovative ways. For example, USA Today has been using its Wibbitz software to create short videos by condensing news articles into a script, stringing images and video footage together, and even adding narration with a synthesized machine voice.

Also, AI can create news and reports for small audiences who follow niche and local topics. By analyzing what is trending across the country, automatic news AI can easily generate relevant news and stories increasing the scope of content provided by the media outlet. Since the automatic generation of content is less labor-intensive, media companies can increase their exposure, and, ultimately, profits. Online media can also become more flexible and dynamic with AI keeping the data in the machine- and human-written stories up-to-date. AI can do real-time monitoring of new events, facts and data, and automatically update published stories with the most recent details. In the same vein, AI storytellers will automate personal reports generation in personal finance reporting, fantasy sports recaps, invoicing, and other online services that deal with structured datasets.

When it comes to social media, it generates huge amounts of opinions, feedbacks, and sentiments that express user attitudes towards political events, brands, companies, and social processes. Social media profiles supply companies with the demographic data, user preferences, interests, education, and occupation, which allows them to target consumers with a relevant content and advertisements.

However, making use of this enormous pool of user data may be hard without automatic software solutions. This is where AI comes to the foreground. Facebook, Twitter, and other popular social media have been using AI bots and crawlers to understand what users are speaking about, and how they respond to advertisements, sponsored content, brands, and other stuff channeled via social networks. For instance, in 2013 Facebook made all content generated by users crawlable by its AI bots. Facebook Graph Search can crawl Facebook posts, comments, status boxes, and anything you share, like, or follow on the network. This data is then fed to Facebook natural language algorithms that try to learn actionable insights from user emotions and responses. AI solutions that analyze user emotions and attitudes underpin the emerging field of sentiment analysis that is getting more attention among companies and brands. For example, in 2017, Open AI company created its Unsupervised Sentiment Neuron AI that can tell the difference between positive and negative Amazon product reviews. The algorithm learned to recognize nuances of customer sentiment by studying thousands of reviews posted by users on the site of the world's top online seller. Businesses expect that sentiment analysis will help them improve user experience, build brand loyalty, and, ultimately, increase sales. Terabytes of big data produced by social media users each day facilitate fine-grained control of social media marketing and content generation to create better user experience and engagement. New AI research is also made into opinion dynamics in social media. AI capable of predicting changes in public opinion are attractive to political parties and campaigners who are already using social networks to reach out to their electoral base with relevant messages, campaigns, and political platforms.

AI-based recommendation systems have become especially important in the content streaming services, such as Netflix. Recommender systems solve the problem of choice, which naturally arises from the user's exposure to millions of videos offered by these services. Also, recommender systems save billions of dollars for these companies, because better content relevance and availability to users makes the massive growth in the actual video catalog unnecessary. Recommender systems simply unearth video content which users would never discover without a little help from artificial intelligence. This is a reason why Netflix saved up to $1 billion in 2015.

Benefits of AI systems for TV broadcasters and OTT services, however, go beyond that. Broadcast and digital media companies see advantages of AI in the real-time broadcasting analytics. The main benefit of cognitive and media analytics offered by such companies as Veritone, Inc. is the analysis of audience exposure, response, and engagement aiming to improve visibility, transparency, and efficiency of commercial advertisements. AI-based advertisement placement systems used in online video platforms such as YouTube, give companies the opportunity to show their ads to the right person, at the right time. Businesses benefit from fewer wasted views and free up advertising space for publishers and other companies. This results in a higher return on marketing expenditures and brings higher yield per view for ad sellers. This technology is now also gaining traction in TV broadcasting for optimization of TV ads in real time.

AI is also revolutionizing Visual Effects (VFX) and animation that play a pivotal role in the commercial movie production. For many decades, animators and VFX experts used motion capture to produce 3D models of human locomotion. Today, state-of-the-art ML algorithms allow simulating human motion by studying large datasets of video clips.

Researchers from the University of Edinburgh and Method Studios recently produced a that system uses Phase-Functioned Neural Networks for Character Control that models motion based on numerous parameters, such as the previous state of the character and geometry of the scene. AI-based VFX systems produce animations and special effects that are more realistic than traditional techniques. Also, AI animation models are more flexible and comprehensive since they include all conceivable types of motion without the need of manual customization.

Besides that, AI methods help directors improve crowd simulation that plays a crucial role in expensive commercial movies. Modern ML techniques allow learning behavior patterns of crowds from real footages recorded from an aerial view. By studying two-dimensional moving trajectories in real crowds, AI systems can evaluate the impact of environment and adjacent agents more precisely than in hand-made animations and motion capture. As a result, we get AI-made crowds that behave in much the same way as real crowds. AI methods improve crowd realism by creating crowd agents who express social and emotional intelligence reflected in the wide range of emotions, gestures, and facial expressions.

Finally, recent breakthroughs in natural language understanding and emotional AI that can read human faces and emotions make it possible to use AI in the production of movie trailers. For example, in 2016 IBM Watson supercomputer was used to produce a 'cognitive' movie trailer or the Fox horror film Morgan. Watson's AI-based system could track subtle changes in human emotions, a tone of voice, and film music to compile a meaningful sequence of film shots that successfully conveyed the emotional atmosphere of the movie. All these innovations improve the quality and efficiency of movie production that will have direct benefits for filmmakers who harnessed AI in their works.

ARCHITECTURE

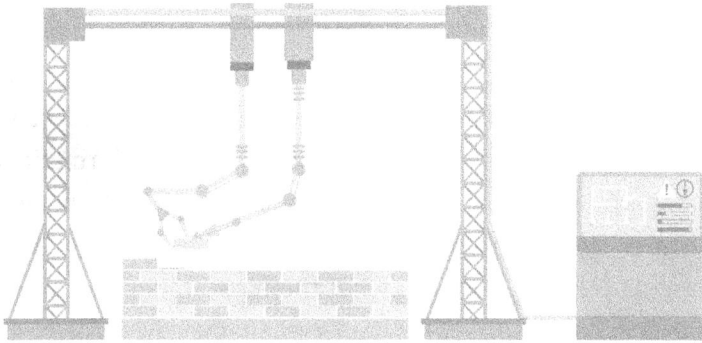

Architects use AI mainly for the computation of complex designs and the provision of design suggestions based on historical design data. As AI technologies become more advanced, AI is gradually entering fields such as urban planning and complete building design. Meanwhile, robotic systems are beginning to apply AI in construction works, an example being automated bricklaying. The smart cities of the future will also require smart buildings and urban planning aligned with the specific needs of a digitalized society in a knowledge-based economy.

A good architectural design requires more than creative and applied thinking. It involves much computation and calculation of forces, material properties, and logistics aspects such as the shortest route to a fire exit. A building is a complex structure and today's designs are based on centuries of experience. Architects use this experience to achieve the highest level of comfort and convenience by using materials in the most cost-effective way. Computer-aided design (CAD) makes all the calculations for them and transforms a sketch into a final design where an architect can only add or remove certain components.

The capabilities of AI go far beyond that, and more advanced design systems are under way. Any building is a data-rich structure. The growing use of connected devices has already provided us with enough information about how people utilize buildings. In the Internet of Things age, an architect can easily gather data on how residents use a building, what the most intensively used parts of a house are, what places people select for rest, and where they place their work desks. Armed with this information, an architect can take advantage of AI and analyze how future buildings can be optimized to deliver maximum comfort. Moreover, AI can provide suggestions about the placement and orientation of the entire building by analyzing historical data about sunny and rainy days, wind conditions, and other factors.

The same holds true in the broader field of urban planning. We now have two overlapping generations that intensely and constantly use mobile devices. People walk the streets, enter shops and administrative buildings, visit entertainment locations, and go to work without ever parting from their smart mobile devices. An AI-powered system can easily track and gather this information, which urban planners and architects can then harness to analyze behavior patterns. Knowledge of these patterns can help them plan better-organized cities and buildings that are more efficient. In other words, AI in architecture is transcending the stage of calculations and entering the realm of intelligent computer design.

It is already possible to feed thousands of old designs into an AI-powered system, and it will provide you with insights into how a building should look depending on its intended use. This approach, however, is based on historical designs, not on current data about how people use the structure. That is why big data is changing the way a future architect will work and create buildings. Selecting materials and calculating forces are relatively routine and easy processes when a computer is at hand. The creation of a completely functional structure is where AI and machine learning can complement the creative thinking of an architect.

A modern urban building is not just a little house on the prairie. It is part of a very complex infrastructure so its space must be utilized very differently from that of a house built 50 years ago. AI can analyze the spatial network of a 50-story building located in the heart of a megacity, providing insights into how to optimize every single square foot. It is not about decorative elements that previous generations of architects liked so much; it is about buildings that utilize all available inner resources and their external connections to other buildings and urban structures.

Furthermore, a system that uses AI for architectural design and features machine learning functionality will evolve. It will enable architects and urban planners to take full advantage of a dynamic design system. Such a platform can learn from past mistakes and successes: it will gain knowledge on which design is applicable in any given scenario, gathering reams of data on building usage in the process. That said, the creative part is to remain a human responsibility for the time being. However, the next generation of AI systems will be able to design a building based on simple input from an architect who wants to create a certain type of building. A step further will be the development of design systems capable of processing commands in natural language. They will work using voice recognition, with architects describing their intended designs and receiving in return at least several basic designs to build upon. Cognitive technologies are developing so fast that we will soon witness beta versions of such AI-powered systems in architecture.

However, the unpredictability of human choices can be a challenge for such AI designs. Sometimes people dislike designs that are the most logical solution for the intended use of a building. Other times they disapprove urban planning that is the most effective solution from a logistics point of view. This is another reason for an AI system in architecture to incorporate machine learning and perhaps cognitive computing technology. There is no rational explanation for people liking gothic cathedrals. Nonetheless, AI can process and assess thousands of past choices and use machine-learning algorithms to provide designs adjusted to the needs of a particular society or culture.

When it comes to creativity and ornamentation (considered one of the main reasons for the appeal of gothic cathedrals), rendering engines achieve what no architect can with pen and paper. Rendering engines are used as a sort of AI in computer games and can create a landscape or a façade in details unimaginable a decade ago. This technology is slowly being adopted in architecture to simulate different landscape and spatial solutions, which required hard manual labor in the past. It allows an architect to simulate various conditions and design solutions adjusted to them. AI can spot any incompatibilities in the design inputs and come up with suggestions for improvement, both in the landscape and the building being designed.

The complexity of a building is no longer an obstacle. Architecture is increasingly using systems that feature Artificial Narrow Intelligence (ANI). They address design problems ranging from materials and forces computation to floor design and infrastructure utilization. On the other hand, AI systems in architecture are far from being creative in the human sense of this word. Still, emerging Artificial General Intelligence (AGI) could change that as well. It's not that important to have creative AI in the field of design. What we need is AI that can design efficient, well-organized, and sustainable buildings.

TELECOM

Telecommunications networks are a perfect field for the use of solutions that rely on AI. Telecom operators deal with millions of subscribers, their networks comprise hundreds of thousands of components, and they deal with a vast amount of technical and business specifications. An average telecom company also has a sophisticated service offering. Carriers grow their networks in scale and complexity on a daily basis, which seems perfect for the use of AI systems and machine learning. Nonetheless, the telecommunications industry is still using only a selection of simple AI tools managing repetitive tasks and pre-defined customer requests.

AI and machine learning technologies are set to enter the telecoms industry in short time. Networks are becoming unmanageable by humans, having millions of users, connections, price plans, and connections to other operators. The Internet of Things (IoT) era will become reality in the coming decades, and then carriers will deal with billions of additional devices connected to their networks. Only an AI-powered system, using sophisticated algorithms, can manage such a complex system. There are automated billing and customer management systems already in place, but they perform only simple tasks and have limited functionality to adjust to the needs of a customer. Implementation of machine learning technologies can allow provision of custom service or billing to any subscriber.

Companies like Afiniti try to provide more sophisticated solutions where AI is used to optimize the work of call centers of telecommunications companies where thousands of calls are received daily. The platform of the AI unicorn that is valued at $1.6 billion is matching caller's phone number with some 100 databases that contain credit history data, purchase data, income data and other information that is analyzed to redirect the call to the most appropriate agent, based on the agent's own work history.

This is, however, only the top of the iceberg. A human cannot deal with the routing and traffic passing through a hardware or virtual network consisting of thousands of components that connect simultaneously to hundreds of thousands of endpoints. Furthermore, it is beyond human ability to predict which route will be the most suitable one at any given moment and how long a usual connection from a user lasts. Predictive analytics can analyze these patterns and select the most optimal connection during a certain period and for a specific user. A select number of carriers have already adopted such technologies although much more is to be desired.

Automated data routing is widespread but we are yet to witness the rise of technologies like self-optimizing networks (SON). A SON network is expected to structure the network without any human intervention to meet pre-designed goals and limits. No human operators, no manual commands – the software tracks the network condition and then takes the necessary measures to provide the best service. This, however, requires overwhelmingly complex algorithms that are still in their early stage of development. Once these algorithms are made usable, the telecoms industry will be able to sustain even higher growth rates. The future lies with Software Defined Networks that will provide the foundation for the development of next generation of intelligent networks.

Neural networks and cognitive computing are the technologies that will provide the backbone of these future networks. At present, the AI systems used across the telecoms industry usually provide suggestions to a human operator who takes the final decision what action the network should take. Machine learning is used to some extent, but the process of learning is largely defined by software developers. With cognitive computing and neural networks, this will change because these technologies can take accurate and independent decisions based on previous events – they are self-learning. A network featuring such capabilities can learn from past mistakes and improve the service continuously, analyzing billions of possibilities in the meantime.

After the IoT network becomes viable, the telecoms will witness their networks grow 10,000-fold. On the other hand, any network, moreover a complex one is experiencing entropy. A network that is not under control and whose components are not managed properly will gradually fall apart. A machine learning algorithm can prevent such events by recognizing patterns of decay and taking preventive actions. We are far from having such an AI system, but cognitive computing is a promising technology that can do the job after some further research and development.

Imagine the complexity of a national carrier's network where millions of home, office, and mobile devices act as endpoints. These endpoints constantly go on and off, transmit and require data. One should add the booming use of cloud services that also require massive bandwidth. The future telecoms network should be extremely dynamic and fault tolerant while load balancing is becoming a critical component of any telecommunications service. A human operator needs visual representation of the state of the network to take reasonable action. Humans read data in the form of charts, tables, or graphs. An AI system deals with data directly; it is not interested in any outside events and thus can understand the core meaning of any bit of data. AI can extract patterns directly from data, not by looking at historical charts or graphs.

Speaking of machine learning, we can say that it is not hard to create a chatbot that converts prospects into customers or provides customer support to subscribers of telecom services. This was possible a decade or more ago but did not happen at that time because the interface was rudimentary and mobile connections were not widespread. What is hard is to replace a human operator with a machine that is capable of learning and taking independent decisions. However, the telecoms industry is among the verticals where overwhelming amounts of data are processed. A single AI system will barely be able to deal with trillions of possibilities regarding routing, the status of connections, the state of devices and so on. With the IoT, these computations are becoming even harder to complete within a reasonably short period i.e. milliseconds. Hence, the future of the AI-powered telecoms network will probably comprise a cluster of AI systems working simultaneously to manage all requests, data traffic, and connections. It should be self-optimizing to get the best results.

Researchers already work on the development of networks that combine reinforcement learning and deep learning to create networking systems that will be able to self-adjust and take meaningful actions depending on experience and pre-defined policies. A deep neural network is using algorithms that utilize many hidden layers to extract information for any component of the networking infrastructure from both the carrier's hardware and connected devices and networks. Then, the algorithm learns how to predict results for pairs that have not occurred before – it may be a router with new software update connecting to a device with brand new

OS version, for example. This deep neural network can track and predict network states and learn over time how to deal with any state and endpoint, even unknown ones.

In the next few years, researchers and telecoms will create more and more software models of telecommunications networks where AI will be implemented to test the ability of the deep learning algorithms. AI will be used to optimize the core infrastructure of the carriers before machine algorithms are applied to manage the networks. Managing telecoms infrastructure is a challenging task but once this goal is achieved the industry will move on to adopt solutions managing entire networks. Gradually, AI will deal not only with customer service but network management across the telecoms industry. It is a trillion-dollar business after all.

LOGISTICS

Movement of goods and raw materials in both domestic and global markets is undergoing a massive transformation, with a growing number of companies across the industry starting to deploy solutions inclusive of AI. Automation and optimization of supply chains are now witnessing unprecedented opportunities, the 2016 Logistics Trend Radar report concludes.

The usage of AI has a broad scope of applications across the logistics industry board. These applications cover fields ranging from manufacturers' delivery chains to warehousing, to last mile delivery. According to the report, the entire industry will undergo massive changes in less than a decade due to the adoption of automated solutions and AI-powered platforms. For example, there are already a dozen of startups developing solutions for global shipping making use of AI to match manufacturers, crop growers, and shippers. The usage of such AI-enabled platforms significantly reduces shipping costs, as well as the overall time required to transport goods both domestically and abroad.

Other unexpected uses of smart devices in logistics exist as well. Smart glasses that were designed with another intended use are now being applied as a tool to speed up warehousing activities, as well as the elimination of manual operations. A warehouse worker can find the required container without having to manually check labels and documents. The same applies to customers, who can now easily find their goods in a large warehouse.

Adoption of robots is another field where logistics will witness major changes. A good number of companies already experiment with collaborative robots that work alongside human employees and eliminate repetitive tasks. Robots are also useful in taking physically demanding tasks from humans in the industry where many such tasks are common. If a self-driving vehicle is also a robot, with some AI capabilities, it is another sphere where logistics will see further changes. This type of vehicles can speed up the entire logistics process and make obsolete some tasks that are currently performed manually.

Some leading companies are investing heavily in the development of self-driving trucks that could one day transport goods autonomously across national borders. Many experts believe that these trucks are already adoptable for domestic use, although more intensive testing and development is still required. There are also certain regulatory and policy changes to be considered before widespread adoption of AI-powered vehicles can begin in earnest.

Nonetheless, drones are already being used for the delivery of food and goods in some locations. Such drones use advanced technology to find the required delivery spot, taking advantage of intelligent image recognition systems and map reading capabilities. The possible use of automated vehicles such as drones, trucks, and forklift trucks is virtually unlimited across the logistics industry. However, the industry is too complex to rely solely on automation but adoption of automated systems is gaining pace and will continue in the long term.

There remains a restraining factor; namely, the unsecured ecosystem of the IoT. Much improvement is still required before it can be widely used for transportation purposes. For example, a home device is already able to communicate with the software of a shipper, but in the most of the cases the communication will lack the required security to protect personal or corporate data. The IoT networks are still hackable, which prevents a wider adoption of devices in the field of shipping of goods and materials. Nevertheless, industry analysts estimate that the logistics industry faces a $1.9 trillion opportunity across the IoT market by 2020. In fact, data-driven logistics is the natural way for the industry to develop in the next few decades.

Data-driven decisions have been the backbone of the transportation industry for centuries, but the usage of AI-powered systems allows shippers to markedly reduce their costs and find the most effective routes for shipping. Transportation of foods from a region in which many manufacturers are located to a large urban center requires a good deal of careful planning. The price of transporting goods to a capital city is not the same as transporting goods to more rural routes. Supply and demand differ from region to region. Hence, carriers cannot demand the same price. The same applies to the availability of shipping orders and the vehicles available to transport them. Therefore, a logistics company can use an AI-powered system to analyze all the data available, before providing a solution based on all factors involved.

Data-driven decisions based on the usage of specifically designed AI solutions will determine the future of any company that is involved in logistics, as well as the supply chain. Look at the following example. A small manufacturer or farmer in Latin America, who is seeking out the most cost-effective and speedy shipping solution and wants to check the status of its shipping from start to finish. It will take hours or days for him to check all available offers from shipping brokers, to select the most suitable one and then to communicate further regarding the delivery of the goods. A software system with AI capabilities, featuring a mobile app, allows him to enter his requirements to gain immediate access to a selection of possible transportation solutions. It is fast, simple to use solution, and is very cost effective.

Moreover, this technology allows the tracking of goods in real time, providing estimated arrival time for goods within a single software platform. The usage of such software systems, which are both mobile and include AI, is still related to the yet fragmented market for mobile services. Transportation companies are still reluctant to adopt mobile solutions because of things like roaming charges, unreliable connection (in some areas), not to mention different standards of mobile services and communication protocols. Security concerns also play a role.

Nonetheless, the future of logistics lies with mobile and AI systems. The beginning of this century saw for the first time in history software systems capable of outperforming human employees regarding decision-making concerning transportation, warehousing, and last mile delivery of goods. Furthermore, automated systems that use artificial intelligence can provide instant solutions to complex problems. No human can know the exact route and status of all ships sailing the oceans at any given time, but an AI system is capable of tracking them all in real time. Thus, the status of any single parcel sent by a customer is readily available at any time.

AI-powered systems are able to cover the entire supply chain procedure from the initial transportation request, to delivery, right to the end – the customer. Machine learning allows AI systems to track as well as predict the availability of goods within a warehouse and then request a delivery of goods, communicating directly with the manufacturer and the transportation company. The same technology, backed by IoT devices, can predict when a farmer in Bolivia will need his bananas transported to Europe or North America. The next-generation logistics company will use a mix of connected devices, mobile technology, and machine learning capabilities of an AI platform capable of taking instant and reasonable logistical decisions.

EDUCATION

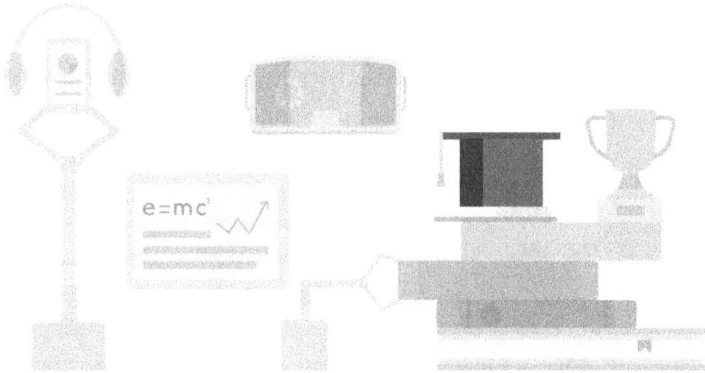

Education is one of the fields where AI has not yet been implemented on a broad scale. One should not make the mistake of not differentiating science-developed AI from the application of AI for educational purposes. On the other hand, education offers tremendous possibilities for the adoption of machine learning, cognitive computing, cloud technologies, and AI technologies as a whole.

No student is the same, and in the past, a teacher would take only a few students in the class to oversee their progress and respond to their specific educational needs. This approach is not possible in a global world in which every student attends some classes, but not all students can learn at the same pace and thus be equally on par in all subjects and school disciplines. Hence, a system that contains AI capabilities can adapt to the specific needs of a particular student and indeed control the entire schooling process to achieve the best results.

This does not mean human teachers will become obsolete in the near future. An AI-enabled system can gather data and use machine learning to provide suggestions to scholars, who in turn stand to gain knowledge about gaps in their teaching and are able to address the particular issue. Such a system can provide insights about individual students that struggle with specific subject matter, whence a teacher can take specific actions to improve the overall learning process.

Creating online platforms that connect students, parents, and teachers alike is only the start of the process. This technology is already in place. Adoption of connected devices, or Internet of Things, introduce far greater possibilities how to monitor and control the entire process of teaching and learning. We speak not of a simple connection between two devices, but of machine learning capabilities that allow the teacher to gain essential insight: when the student is most actively engaged in his or her homework, and how much concentration time the student is capable of and so on. Online tutoring is gaining pace, and education needs enhanced AI systems to assist the teacher in certain aspects, including capturing the interest of the student, when in the past their attention span might have been negligent.

An AI-based educational platform can track how a student is answering certain questions and then guide the learner through the best paths to achieve better results. After all, not all answers are simply incorrect.

Machine learning technology can masterfully guide the student through different paths to success. Furthermore, intelligent software systems can gather information on how all students are coping with the curriculum and then to provide suggestions on how the specific curriculum needs to be corrected.

Another option is for a system to incorporate crowd-sourced teaching capabilities where the AI can compare the tutoring and learning capabilities of all the participants and in turn match the best pairs of tutors and students. The entire learning process can be improved during this comparison. A relatively simple educational AI system can easily find low-quality teaching content and provide links only to the suitable curriculum.

Within a decade (or less) we should acquire a fully personalized content for each student. The problem with mass education is that each textbook is invariably written with the average student in mind. Machine learning technology can differentiate between the students and their individual capabilities providing a completely tailored content to every student individually. Moreover, any school or university can use a mobile app where the student can share his or her advances, the data of which can be analyzed by a complex AI system. Custom textbooks offer a great opportunity for the improvement of the entire teaching process and should achieve results that will produce a generation of well-educated people that are not lacking skills because of, say, a lack of learning speed in the subject of mathematics.

Recently, some experts introduced the concept of lifelong learning companions that are accessible from any device anywhere. This can fulfill a dream of generations of teachers and students alike because such an AI platform would provide all the required knowledge at the appropriate time about the individual concerned. Such a personal, powerful assistant would gather piles of data about its students, which would enable it to predict their educational needs using profound learning technology. For example, a personal AI companion would provide one with information that one would otherwise have no choice but memorize. You cannot remember all the facts you studied during your secondary or tertiary education, but a personal assistant can analyze precisely the facts and data one needs, while suggesting one pays attention to them.

Sure, there are some issues yet to be solved regarding AI-assisted learning. There is no machine that can show human-like empathy or collaborate on an actual human level. Some of the next generations will have to deal with this problem until the current AI technologies are developed to a stage at which computer-assisted learning will be more fun to use. Virtual learning environments (VLE) eliminate much of the genuine human interaction that is also a vital part of any education. Therefore, the appropriate balance between AI-powered systems and teaching by humans needs to be constituted in the not-so-distant future. The technology behind AI tutors should make a further step in such a direction before they are able to stimulate creativity that is beyond basic learning.

Nonetheless, AI programs are very useful in providing feedback to both teachers and students since they can monitor, collect, and process amounts of data no human can process. Data-driven educational solutions will become more and more widespread in the next decades before they evolve into more complex systems serving a greater variety of needs. Personalization of content provided by intelligent systems is yet to reach maturity in educational terms. Current technologies used by search engines, for example, filter through a certain amount of information based on your location, previous searches, etc. Thus, a very complex algorithm is required to adapt to the needs of a student without preventing him or her from obtaining important information.

To conclude, the entire process of education is undergoing a process of massive change in which students can learn from anywhere at any time. Such mobility will fuel the growth of AI systems adoption across all educational organizations and will boost further mobility and independent learning. As in other fields, teaching robots and AI systems will never replace face-to-face learning, but will inevitably create a new learning ecosystem in which machines will deal with many tasks now performed by human teachers.

INSURANCE

The insurance industry is one of the most conservative when it comes to adoption of new technologies, but artificial intelligence is slowly changing the way the insurance business is working. Some of the possible applications of AI algorithms across the insurance industry include identification of emerging risks related to climate, economy, demographics, and overall insurability.

The use of AI-powered systems to analyze observable trends is another field where intelligent systems can provide invaluable insights. Insurers adopting AI solutions can more easily develop new coverage products to respond to emerging trends and risks. Systems featuring machine-learning capabilities gradually replace humans in collecting, cleansing, and synthesizing vast amounts of structured and unstructured data enabling insurers to focus their efforts on the development of new business models and products, instead of dealing with data gathering and data processing.

Adoption of AI solutions across the insurance industry will influence efficiency and automation of current customer-facing, underwriting and claims processes, according to research by PwC. Nonetheless, this is only the start of a continuous process where AI systems will enter fields like identification and assessment of emerging risks. Over time, software featuring machine learning and AI capabilities will start to identify potential revenue sources. By assessing trends and emerging risks, an AI system can provide insights for individuals, companies, and lines of business thus greatly improving the overall profitability and risk mitigation.

Machine learning and AI technologies are set to change the entire business process a traditional insurance company is following for centuries. The traditional operational process of an insurance company starts with the identification of a pool of customers and assessing risks associated with them. Then the insurer will target those customers and will evaluate the risk for each class of customer. Eventually, the insurance company will sell policies that spread the risk over the pool of customers by offering differentiated policy price for each class of customers. The next step of the business process is to retain as many clients as possible by lowering prices for long-term contracts.

How are AI and machine learning are disrupting this traditional model? Identifying, assessing, attracting, and retaining a customer requires significant efforts regarding both data collection and tracking these data over long periods. Insurers pile large amounts of data on individual customers, groups of customers, and entire markets. However, those data are mostly unstructured or semi-structured, and insurers perform evaluation using old-fashioned models that include a great deal of manual work. Even the simplest AI system can filter and organize the collected data faster than an agent and also provides error-free results. Moreover, the use of AI software to analyze the collected data allows insurance agents to focus on their core duties of closing deals and retaining clients instead of pouring through piles of unstructured data. Thus, the AI platform does all the hard work of identifying prospects and assessing risks while the agent deals with the very process of closing deals.

As we said, insurers possess vast amounts of data on individual and corporate clients and prospect customers. This information, processed and analyzed using machine learning algorithms, can be used to increase customer engagement and retention rates. Customizing an insurance policy offer is a difficult task where an agent must assess and consider numerous factors and historical information. An AI system featuring machine-learning capabilities can provide tailor-made suggestions for each specific client, be it individual or corporate one. Furthermore, RPA (robotic process automation) eliminates many manual tasks performed by agents in the past, which in turn boosts efficiency and reduces costs.

Whatever data an insurer possesses on a prospect or a client, sometimes it is hard to assess the possible risks that the client bears. A specifically designed AI system can process and analyze these granular data and then provide insights whether the customer is uninsurable because of excessive risks. It can also track the respective data over time and flag the customer as insurable again once certain conditions change. Also, an automated insurance agent could contact those customers that are insurable again and propose custom offers to them.

A viable robo-insurer is still a matter of further development, but many companies already work on the design of such platforms based on current insurance bots. On the customer side, today's insurance bots are only able to provide suggestions, and compare policy offers based on information input by a person. The next generation of robo-insurers, powered by AI, will be able to give you a fair policy offer based on your historical, personal data, habits, and preferences tracked over time. Comparing offers is the simplest thing – providing a suggestion that is the best fit is a matter of machine learning and intelligence. For instance, sometimes the best fit does not depend on the price of the insurance policy but factors like additional risks covered, long-term benefits, and a set of personal preferences.

Claim investigation is another insurance process that requires a lot of effort from insurers. Agents spend a lot of time assessing claims and validating them, which is a process that an AI system can complete in seconds. We speak of more simple claims where certain patterns can be followed and where an algorithm can identify all valid claims and earmark others that need further validations.

Likewise, fraud prevention has always been a pain in the arse for any insurer. Machine learning algorithms, coupled with AI, can easily spot most fraud attempts and analyze patterns to predict how these attempts could develop. Thus, an insurer can be alarmed not only about actual frauds but also about attempted fraud in which new deception methods are used. Predictive analytics is a powerful tool to detect such cases.

AI is also able to improve the overall cost structure of the insurance industry by enhancing all the components of the insurance profit structure. The four major factors an insurer should have in mind regarding the profit structure are premium earned and the investment income as opposed to underwriting cost and claim expenses. Application of AI systems can introduce improvements within any of these four components of the profit structure.

As a final point, AI systems and machine learning algorithms can optimize and enhance almost any process or procedure within the insurance industry. Both insurers and customers, however, should adapt to a new mindset with less paperwork within a fully digitized environment. This will be a relatively slow process because of the very nature of the insurance business where a great number of factors determine both the insurer's decisions and the customer's choice of insurer.

PART 2:
HOW DO BUSINESSES
TRANSITION?

This section explains the basic principles businesses should follow to transition into a new world of artificial intelligence

It's important to understand why AI is a much-needed transformation, and it's also crucial to acknowledge why it's important for your business. In the current world of fast-paced innovation, companies are facing an extreme lack of acknowledgment what is AI and how it can help.

An effective AI strategy is to understand the benefits of machine learning and respective applications within the company. It is a fact that most industries simply don't get what is AI/Deep Learning and other buzzwords.

And it is understandable, just from looking at the chart of several publications on the AI domain, one can't simply grasp all the knowledge, and that's ok. Most of the thousands of machine learning experts may have no idea what new technology is out on the market, thus making the knowledge about capabilities of relevant technology very scarce.

GROWING AMOUNT OF AI PAPERS SUBMITTED

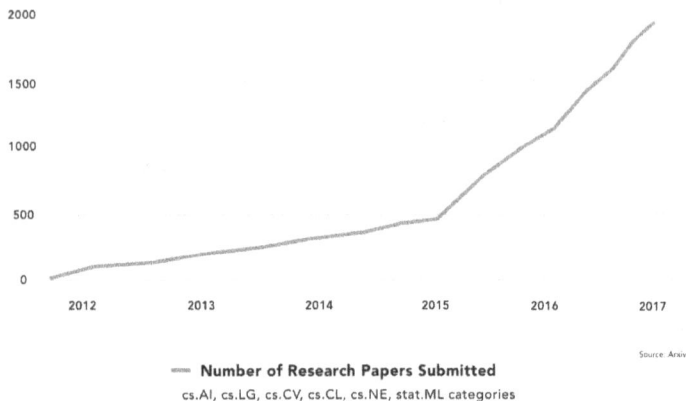

Source: Arxiv

— **Number of Research Papers Submitted**
cs.AI, cs.LG, cs.CV, cs.CL, cs.NE, stat.ML categories

March of 2017 saw almost 2,000
submissions in these categories

The concentration of the knowledge resides within two big chunks, the corporate world of big giants such as Google and academia. And of course, there is a small portion of getaway talent working on startups that broke through those two worlds and decided to go on their own.

GROWING USE OF DEEP LEARNING AT GOOGLE

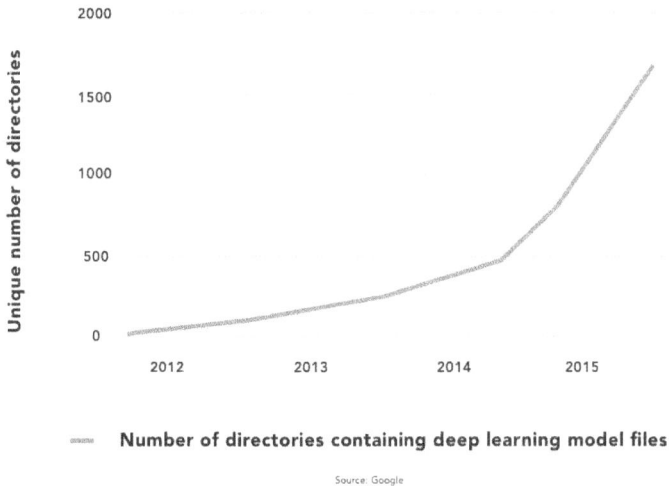

Number of directories containing deep learning model files

Source: Google

There are two ways companies can transition into the brave new world of machine learning. The first one is to silently wait until technology specific to your business case is made widely available and democratized for the mass use. The second way is to take a hands-on approach and create that technology yourself.

There are good and bad reasons to begin using AI in a company. A good reason would be the real product need for it; the bad reasons range from building up PR buzz and just being on top of the cool trendy word. AI is not a fancy optional thing you can try out without investing significant time and resources. AI requires dedication from whole organization and often restructuring around the use of machine learning.

CHAPTER FOUR

EDUCATION

This section covers the potential applications and limitations of Machine Learning.

TYPES OF LEARNING

You don't need to learn the full history of neural networks or understand how the deep learning works inside the box, but you should understand what exact technology you are looking for and what are the business outcomes of its usage.

Since the core value of AI is the ability to learn, it's mandatory to understand what types of learning exist.

Researchers divide machine learning algorithms into general groups:

- Supervised learning

- Unsupervised learning

- Semi-supervised learning

- Transfer learning

- Self-taught learning

- Reinforcement learning

Supervised Learning

The reason why it's called supervised is that the process of the algorithm learning is guided through the data it's provided to learn on. The data that is consumed by the algorithm is labeled beforehand, thus providing the necessary expertise on what is what in the dataset.

The learning here happens in iterations; each iteration algorithm makes a prediction which is corrected by the data provided by the teacher. The process repeats itself until algorithm reaches the minimum required quality of results.

The most common example of supervised learning is the regression, where the output of the algorithm is a real value such as time, money or any resource. The term used to describe this type of output is called continuous. The idea behind regression is quite simple, given the statistical data you can observe the relationship between your target output and input parameters. Then you can take this learned relationship and map it to the real observed data.

The example of regression is typically explained with the real estate pricing prediction problem.

The input can be parameters such as city area, number of bedrooms, year of construction, etc., and the output is the price of a house matching those parameters. The more input (affecting) parameters you choose - the more complex the learned model will become.

There are also many types of regressions which are used in relation to the specifics of the machine learning problem. The simplest form of regression is the logistic regression; it fits to the problems where the output is binary meaning it belongs to two types of categories, like black or white, has diabetes or does not.

Classification is the more complex version of logistic regression where the output of the machine learning algorithm lies between a set of categories, meaning a discrete output.

An example of such classification problem would a be an algorithm that analyzes pictures of pets and classifies them into separate categories such as dogs, cats, or rats. The most common use of classification that we see every day is email spam filtering and email grouping that Gmail does for its users. Emails are now automatically classified into the primary inbox, social, or promotions.

Unsupervised Learning

In contrast to having a teacher and a dataset that has clearly labeled data, unsupervised learning is a type of machine learning that is done with the absence of teacher or additional labeled information.

The main purpose of unsupervised learning is the ability to find hidden structures or relationship between data that no teacher can spot or has yet to produce the labeled dataset. Simply put, the goal of unsupervised learning is to learn more about the data itself.

The most popular problems solved by unsupervised learning are clustering and association.

Clustering is a problem where you need to discover the inherent groupings in the data. A real-world example of clustering would be grouping customers by what they are buying.

Anomaly detection is another interesting application of unsupervised learning that is especially useful to industries where timely anomaly detection is a crucial mechanism to prevent fraud.

Semi-supervised Learning

A semi-supervised approach to learning is the way to use a small amount of labeled data with a large amount of unlabeled data. Researchers have found that this approach of using unlabeled data in combination with a small amount of labeled data can result in a magnitude of improvements in learning accuracy.

Many of the real-world machine learning applications fall into this area simply because most of the time it's very expensive or time-consuming to acquire fully-labeled datasets when the unlabeled data is quite cheap to obtain.

Transfer Learning

Different machine learning problems have many tasks that share some common similarities. To get the benefits of this similarity, researchers introduced the notion of transfer learning that can focus on storing knowledge gained while solving one type of a problem and reusing that knowledge for a different but related problem.

One example is a neural network that learned to recognize cars can be applied to the problem of recognizing trucks, since most of the features like wheels, lights, windows are quite similar and can be shared across those two problems.

Self-taught Learning

In the types of problems where labeled data is extremely hard to obtain, there must be some way to bootstrap learning accurately without much of the labeled data. One approach presented by Stanford researchers is self-taught learning.

This type of learning is based on the idea of using unlabeled data, such as randomly downloaded images from the Internet, to make algorithms learn very high-level concepts such as edges and other basic visual patterns.

The hypothesis is that these high-level visual patterns that are self-learned from the random images can be applied to the specific problem, such as classifying elephants and rhinos on the pictures.

The benefit is obvious; it's quite hard to obtain large datasets of such images thus the ability to learn most of the needed basic visual concepts can be accomplished through unlabeled images which most likely will have needed basic features.

Reinforcement Learning

There is a subset of machine learning problems which doesn't fall into either of the described categories. The problems where the algorithms operate on a notion of making good or bad choices/actions and being rewarded /punished for those is called reinforcement learning.

It's like unsupervised learning since the algorithm doesn't have access to the predefined rules and doesn't know what is good or bad. This type of learning is an attempt to mimic the behavior of biological entities, a car could be the agent and the goal for it would be to find the best way to drive from one place to the other without hitting into obstacles.

The important part of reinforcement learning is the presence of a reward that indicates if some action or set of actions was positive and punishment that lowers the score and indicates that certain action was not positive.

NEURAL NETWORKS

If we have so many algorithms for all types of learning, why exactly are neural networks getting all the attention lately? In comparison to classical supervised problems such as regression, neural networks allow you to build out a more complex non-linear relationship between given inputs.

As a simple example, you can treat neural networks to map out all those numerous relationships between chess play figures on a board, instead of just being able to evaluate how good the position of your king is to all other figures.

Neural networks can be applied to both supervised and unsupervised learning. The difference is again the presence of a teacher with the already known definition of results. Neural networks in a supervised environment are used for classification, pattern recognition tasks, function approximation and prediction systems.

Neural networks in an unsupervised environment are often used to learn more about the input like the clustering problem neural network learns formalized representations of the input. For example, you can feed the network huge amount of text documents and it can learn a simplified way to describe those articles so that you can easily find similar documents, or similar content of these documents.

This is usually achieved through autoencoders, models that learn to reconstruct the original input from a smaller amount of data with the desirable amount of reconstruction error. In contrast to clustering, it doesn't group inputs but rather creates meaningful representations out of those so that they can be used for clustering.

IMAGE RECOGNITION

VISUAL INPUT OF THE NEURAL NETWORK

```
0   0   0   0   0   0   0   0   0   0   0   0   0   0   0   0   0  0
0   0   0   0   1  12   0  11  39 137  37   0 152 147  84   0   0  0
0   0   1   0   0   0  41 160 253 255 228 162 255 238 206  11  13  0
0   0   0  16   9   9 150 253  45  21 184 159 154 255 234  40   0  0
10  0   0   0   0   0 120 146   3  10   0  11 124 253 255 107   0  0
0   0   3   0   4  15 236 216   0   0  38 109 247 240 169   0  11  0
1   0   2   0   0   0 253 253  21  66 234 241 255 164   0   5   0  0
6   0   0   4   0   3 252 250 228 255 255 234 120  21   0   2  17  0
0   0   1   4   0  21 255 253 251 255 172  21   8   0   1   0   0  0
0   0   4   0 163 225 251 255 229 120   0   0   0   0   0  11   0  0
0   0  21 162 255 255 255 255 126   6   0  11  14   6   0   0   0  0
3  70 242 255 141  66 255 245 189   7   8   0   0   5   0   0   0  0
26 221 237  98   0  67 251 255 141   0   8   0   0   7   0   0  11  0
125 225 141  0  87 244 255 208   3   0   0  13   0   1   0   1   0  0
145 248 228 116 235 255 141  34   0  11   0   1   0   0   0   1   3  0
85 237 253 246 255 208  21   1   0   1   0   0   6   2   4   0   0  0
6  23 112 157 114  32   0   0   0   0   2   0   8   0   7   0   0  0
0   0   0   0   0   0   0   0   0   0   0   0   0   0   0   0   0  0
```

Handwritten digit "8" represented as color intensity values (left) and pixels (right)

Think of how hard it is for you to spot a cat in the image; it's not that hard since our brains are used to these types of tasks and spot the objects we seek for right away. It's even baked in our evolutionary mechanisms of detecting predators and other harmful objects.

But it's not an easy task for machines since the only thing they see is raw pixels unless they are trained specifically to detect something.

The field of computer vision has benefited greatly from the expansion of open-source, deep learning technology and there are a couple of industry popular models that are to thank for most of the image recognition advances recently. They often surpass human accuracy in certain tasks.

One of these models is called Inception, and it is trained to recognize over 1000 categories of images. Usually, these models are trained on the public datasets such as ImageNet.

The architecture of this deep neural network is super complex and makes it impossible to understand if you are not a deep learning researcher so this part doesn't bring much interest here. What's interesting is the end business results that this type of machine learning can produce for companies of any scale.

INCEPTION MODEL ARCHITECTURE
Partial example of deep learning model used for computer vision applications
Source: Arxiv

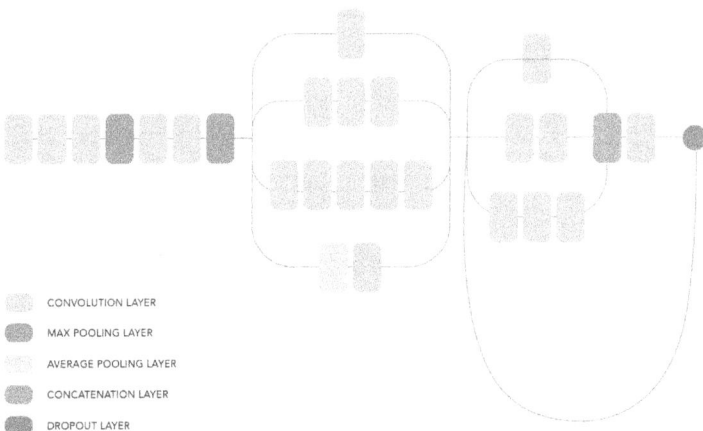

CONVOLUTION LAYER

MAX POOLING LAYER

AVERAGE POOLING LAYER

CONCATENATION LAYER

DROPOUT LAYER

So, what is the actual input for those neural networks? In a case of image recognition, it's the raw pixels. Each image can be represented as a set of pixel values: their coordinates and brightness, since object recognition usually doesn't take into account colors unless the task is specifically related to color recognition.

FACIAL RECOGNITION

One of the examples of such image recognition use is the ability to detect faces in images or videos. It has been in use since early days of social networks allowing people to tag their friends in photos by highlighting the rectangle areas around their faces.

Every smartphone has a folder of images classified as "selfies" in their photo apps which are formed by scanning through your pictures and classifying them.

More traditional businesses such as Disney are adopting facial recognition to help visitors find all those photos that they take on board a cruise liner. The technology automatically tags people in the pictures which allows it to create designated albums for each person on board. This process was previously completely a manual task using paper notes and numbers associated with the current sequence of photos. Royal Caribbean Cruises have also revealed its goal to get into the tech space with using mobile app assistant powered by facial recognition to allow check-ins at ports and automatically unlocking cabin doors.

There are also uses of facial recognition across education institutions to check the student presence in class during tests and for online exams using a student's webcam to verify identity and prevent cheating.

The biggest use of this facial recognition technology is security, from personal security to enterprise security and even community safety.

The most common example of facial recognition use in the consumer market is smart doorbells with smart home security cameras. Use cases range from letting owners into the house to confirming that it's someone you know at your front door or a random stranger you have never seen.

There is also a huge market for solutions aiming to recognize emotions. Companies like Bentley used facial expression recognition in the marketing campaigns to suggest car model types based on emotive responses to certain stimuli. It is also used to help measure

TV ratings and create new virtual reality experiences.

Emotion detection and facial recognition are also a huge buzz in retail. These capabilities provide opportunities to companies to identify the emotions customers felt as they walked through the store thus creating a truly personalized experience. Even though most customers keep a straight face when shopping, the new software by Emotient can register micro-expressions–tiny flickers of emotion that show up on faces before they even know they have registered an emotion or are able to control it.

Fujitsu recently began distributing licenses for retail facial recognition technology capable of recognizing demographics of individuals. This feature itself allows companies to optimize their advertising efforts around the estimated age and gender. The application of a such narrow targeting would greatly increase the relevancy and minimize costs of these marketing efforts.

MODERATION

With thousands of users creating content every minute, it is difficult for companies to always catch inappropriate content being added. Be it photos or images, you can always find something that is not safe for work or simply violates guidelines.

It doesn't make sense for companies to use resources to moderate this information on a scale so these things are automated with special NSFW-types of models that easily detect unwanted content and can protect the brand from misleading exposure.

GENERAL OBJECT DETECTION

Pinterest's visual search engine is a perfect example of object detection in a real-world application. Pinterest visual search tool lets users zoom in on a specific object in the image and discover visually similar objects, colors, patterns and more. For example, you can zoom into a lamp and will automatically get hundreds of similar lamps in search results.

In different applications, object detection can range from detecting only colors to detecting specific types of objects or even their relationship with other objects.

And most likely you've already seen AI nutrition apps that can estimate how many calories you consume by simply taking a picture of your lunch. This usually works with general object detection or more custom models.

There are also several use cases when the objects in the images or videos are recognized not for what they are but what brands they showcase or which companies sell them. This is useful for more general type of applications such as shopping lists.

IMAGE GENERATION

Synthesis of images is another useful application of generative neural networks that learn to emulate the type of images they've been learned on. This type of generation can be used for a wide range of applications, from auto completion of corrupted images or fragments of photos to generating completely new images based on the trained model.

One such application is artistic neural style transferring that has received a lot of attention recently and created a couple of new startups. The concept behind this technique is the ability to generate or transform existing photos with the style of another image.

Prisma is an app that allows you to transfer the style of one image like a Picasso painting onto another image, like your selfie photo. These style-transferred photos don't look realistic of course but instead they look like they have been painted. Facebook announced similar photo and video filters coming to its mobile app too.

AUDIO RECOGNITION

AUDIO INPUT OF THE NEURAL NETWORK

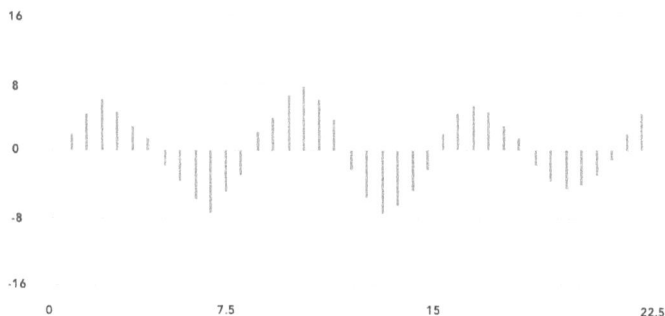

With audio recognition, the input is a more complex structure, since audio is not an image and can't be represented as pixel positions and their brightness. Audio is a one-dimensional wave so at any given time there is just one value that is equal to the height of the wave.

Nature sounds, musical instrument playing, or someone saying "hello", will all be a wave encoded as a set of height values. This process is called sampling.

Some experienced car mechanics can spot issues with the brakes or car engines with their ears. These human experts can detect common problems with machines in the respective domains just by listening to the sounds they make. This is a typical anomaly detection problem that can be solved if you have a respective dataset of these common issues.

Predictive maintenance is a large market and there are several companies trying to help solve issues such as saving time, money and even lives. Companies like 3D Signals specialize in detection of acoustical anomalies in equipment. Their acoustic monitoring and deep learning technology monitors sensory data from production line machinery, identifies anomalies, classifies patterns of equipment failure, and predicts issues before they interrupt production.

Companies like OtoSense provide software platforms that enable machines to make sense from sounds in a more general way, giving you the ability to train your devices to spot things you tell them to spot. Any number of physical or virtual devices can be deployed and work together to detect events and describe them, constantly learning from humans and environment to improve.

This is not limited to just factory machinery but can also spot voices, alarms, gunshots, ground or airborne vehicles, and steps. Basically, you can search for and act on any sound event. It even allows you to upload specific sounds that you want detected in your environment.

Security has its benefits from sound recognition. The ShotSpotter system, for example, is used in several major cities across America. It uses highly sensitive microphones around an area to pick up sounds from the street that might be gunfire, and uses the sensors to calculate the location of a gun crime. It then sends the information to the law enforcement to react.

The ShotSpotter system can identify the weapon type and the number of shooters, and makes this information available to police officers on their phone where they can view the gunfire location on their map and even listen to the audio clip of the specific gunshot.

Home security systems such as Audio Analytic help customers make informed decisions based on what happens in their home. Their system can detect things like baby cry, smoke alarm sounds a window breaking.

Understanding various ambient sounds is more complex than understanding speech since speech is quite limited to the number of phonemes, languages and our vocal cords in contrast to more vague sounds like someone breaking into your house.

The most common use of sound recognition is attributed to speech recognition. Things like Siri, Google Home and Alexa have only recently become capable of serving real needs but they are perfect examples of speech recognition in our everyday lives.

General speech recognition task is also a difficult problem since you must overcome such audio challenges as bad quality sensors/mics, noises, reverberation and echoing, and even accents. Each of these things must be accounted for when you are building your own solution.

To build something on a scale of Siri and Alexa, you would need to have unrealistic amounts of training data that can be acquired only by such giants as Apple and Google. That's exactly the reason why Google records and stores every little thing that you say through its speech interfaces. This data is accumulated and is used to further improve the accuracy of the overall system since nothing beats the audio recorded in a real-world environment within the real-life situation like you trying to find something on the Internet.

More vertical companies are working on analyzing more niche types of speech. Take Chorus.AI for example, the company that listens to sales calls and help companies close more deals.

The value proposition behind Chorus is an ability to extract insights from the calls. Chorus simply joins conference calls as an additional listener and records content to be transcribed in real-time. The platform flags important action items and topics that come up during the calls.

The platform they've built not only transcribes sales calls to reveal insights but also makes it super easy to create effective training programs that lead to better sales reps.

This ability is based on their extensive speech analysis and ability to pinpoint common objections or competitor mentions.

SOUND SYNTHESIS

Speech synthesis or so-called text-to-speech is a widely-used technology and it has been getting exponential improvements to both speed of voice generation and realistic sounding of it.

One recent major advances is DeepVoice technology developed by Chinese search giant Baidu. A research team at Baidu presented technology for a high-quality text-to-speech which required far less computational resources and time to generate than previously released WaveNet developed by Google.

Researchers achieved 400x magnitude improvement by making speech generation real-time, since previous implementations took minutes to generate natural sounding pieces of speech.

These major advances are all attributed and spiking up because of the deep learning nature of the algorithms behind them. With the abundance of computational resources, companies like Adobe are even able to create photoshop for voice. Their recent presentation of VoCo product that allowed to edit human speech within a simple typing interface has been all over the press.

The technology behind VoCo can essentially make you say things that you've never said by generating words using speaker's voice. Basically, the algorithms can understand what makes up the specific person's voice and replicate it. This requires a substantial amount of data on this person talking - approximately 20 minutes of real speech.

Essentially technology can be used similarly to how Photoshop is used to make small edits here and there whenever the person made a mistake or said something that would be better changed.

Another company named Lyrebird recently unveiled its first product: a set of algorithms the company claims can clone anyone's voice by listening to just a single minute of sample audio. Besides the ability to clone anyone's voice, it also has an ability to shift emotional cadence of the speech, which has not been accomplished by anyone else so far.

Even though technology can be used to misguide people and biometric security systems, it's still a huge advancement and will be used in many areas effectively. The uses can range from reading audiobooks with famous voices to synthesis of speech for connected devices of any kind to speech synthesis technology enabling people with disabilities speak again. It can even be used for animation movies or by video game studios.

Music synthesis is also another hot topic since it involves such an unpredictable human parameter as creativity and subjectiveness of produced music. There are many open source projects in this niche that achieve quite an impressive result. For example, Google's Magenta is an open source project that is used to generate music. For now, Magenta is quite limited and can produce only one stream of notes (one instrument) but already generates pieces of music that are easy to listen to. There are other projects like DeepJazz that can generate jazz music and BachBot whose goal is to generate music indistinguishable from pieces written by Bach.

TEXT ANALYSIS

Text mining or natural language text analysis is gradually transforming how businesses operate as it provides a set of crucial effective methods to deal with the most popular way companies save data: text. Text attributes to approximately 80% of the stored information for companies worldwide.

Several distinct uses have harnessed the power of text analysis to provide more powerful business operations. The most popular are again tied to the big data's famous search for insights. The ability of smart algorithms to analyze the vast amount of natural language format data to identify hidden trends is the holy grail of big data analytics. The detection and identification of concepts that are commonly appearing in documents is also a very powerful trend spotting mechanism.

Another use of text analysis is document classification that allows you to group and classify documents or its internal content into predefined categories or classes. When there are no existing predefined categories, then clustering or grouping is used.

Information retrieval is used by companies to obtain documents or information from those documents in response to a query. Several companies are working on simplifying access to huge databases by transforming these queries into human language. The most obvious example of information retrieval is any search engine that can find the most relevant results to your query.

Information extraction makes it easier to extract needed information or entities from cluttered sources in a pre-specified, much more readable format. An example of this is Google's ability to show advanced search results with extended information through its knowledge graph technology.

As the problem of information overload is growing daily, the promise of summarization technology is making a huge impact in both consumer and corporate industries.

The app called Summize has been top paid app for some time by providing a way to summarize textbook pages into short meaningful structured pieces that are super easy to digest. It uses a combination of optical character recognition and the ability to summarize knowledge and structure it through extracted points.

Even though summarization may sound as if it's an easy problem to tackle, it isn't. Here are some factors that make it a rather complex task:

- How abstract summary should be

- How many documents should be considered

- Who will read the summary

- How will the summary be generated

- Which genre/style summary should have

There are two types of summarization, the abstractive and extractive. The main difference is if the summary is built out of the new words or sentences or simply from the existing extracted sentences.

Since rewriting may require high-level skills that only expert systems or human experts have, most of the general services only offer extractive summarization. Simulation of these human expert skills requires a combination of natural language generation and extensive knowledge of the domain.

It's important to understand who is the audience that intends to read this summary and what is their goal. If the goal is to simply get an overview of a new topic, then the summary can be quite vague and indicative. This is often the case for analysts who need to quickly skim over the summaries to get the summarized news or events.

If the goal is to deeply understand the topic and make educated choices based on the content, then the summary has to be informative and cover all the details mentioned in the original text.

The number of separate documents needed for the summary is an important metric. When the summary is compiled from multiple documents, the system must deal with repetitive facts and concepts. That's where it gets tricky and may again require a deep understanding of domain and context.

There are even use cases where machines can teach us how to write better. Grammarly is a company that analyzes your writing and suggests improvements. The app is used and licensed by hundreds of companies and universities in the US and it is much more than just a grammar checker.

It can spot repetitive words, overused phrases and even words that non-native writers commonly misuse. The app can select genre options so that suggestions will be in line with the style of the person's writing whether it's professional business document or a casual blog post.

CHATBOTS

A more traditional text processing and conversation building that has been all over the press in the recent years is conversational technology labeled as chatbots.

The convergence of trends such as nationwide adoption of messenger apps and improvements in natural language processing have made chatbots an ideal communication instrument for businesses.

The main reason why this instrument is so effective is its ability to communicate individually and contextually on a one to many basis.

But, of course, the basis for such communication is still text, be it SMS chatbot or a Facebook messenger bot, it still sends you text with occasional media attachments in response to your input.

There are many advantages of using chatbots, most of the prominent are:

- **Availability** – a bot, unlike human communicator, can be available 24 hours 7 days a week without any interruptions

- **Faster response** – a bot can analyze requests for information and retrieve it in under one second in contrast to humans, especially if combined with machine learning algorithms to better understand those requests

- **Personalization on a scale** – bots can way faster retrieve your previous history and tailor responses based on that

- **Costs** – using chatbots to handle basic communication and free up human resources

The main distribution channels for bots are messaging platforms. Currently most of the messaging apps such as Facebook Messenger, Slack, KIK, Skype, Line, WeChat, Telegram support bots to some extent.

The building process of chatbot from scratch is quite complicated, therefore bot frameworks and platforms exist to make the life of developers easier since they don't have to deal with raw nature of machine learning algorithms and natural language processing.

TIME-SERIES ANALYSIS

Time series is a data that has a valuable time order that can't be dismissed when trying to understand this data. Examples of time series are heights of ocean tides, the daily closing value of the Dow Jones Industrial Average, birthrate of the community, and weather conditions.

Time series analysis is used to better understand and model the time-related properties of the data to use it for forecasting - predicting the future outcomes based on the previous history.

For example, the typical use case for time series example would be the analysis of how a specific stock performs on the market. The time series data would be all the closing prices for the stock from each day for the year given.

When working with time series data there are several distinct components that should be carefully described:

- Trend - the increase or decrease in the series over a period of time

- Cycle fluctuation - repeating up and down movements

- Season variation - regular pattern of repeated fluctuations

- Residual fluctuations - random variations or unforeseen causes

This way you can model and predict trends, see if the data shows any seasonality to determine if it goes through peaks and valleys at regular times each year.

You could model it against other related variables such as the growth of a job market or the unemployment rate. Using the information from both time series datasets, you can observe patterns in situations exhibiting dependency between the data points and the chosen variable.

Overall, the field of time series analysis is very old but the recent advances in deep learning and neural networks has brought a lot of powerful methods to apply to this problem. One perfect example is the use of recurrent neural networks.

A recurrent network has connections between its neurons like traditional neural networks, but with the addition of loops and feedback mechanisms that allow the ability to store some form of memory.

This type of internal memory allows neural networks to make decisions based on the new incoming information and previous results stored in memory, which makes it especially efficient in problems related to time series type of data.

Time series forecasting can also be used for anomaly detection. Since a recurrent neural network understands the structure of the given time series data well enough, it can detect if the new data presents fraud or an anomaly in a timely manner.

WHY DATA IS THE KEY TO SUCCESS?

This section explains why data is the most important asset when trying to use artificial intelligence

The volume of available and generated data has grown exponentially in recent years, but not everyone understands that this data becomes the hidden enabler behind many powerful business disruptions. So why exactly it will be the most powerful business asset for the next decade?

Think of data as gasoline for machine learning. As described before, AI algorithms are not manually handcrafted sets of rules that machines follow. They learn through experiences and the more experience we feed into these algorithms, the more accurate output we will have.

In general, types of machine learning such as deep neural networks or deep learning are very "data-hungry".

These algorithms have a certain threshold of dataset size that you need to hit to be able to actually tune dozens of inner parameters in order to come up with somewhat generalizable models.

Recently a number of statements from industry leaders suggest that having better and bigger datasets often helps more than having the most efficient algorithms.

Such a statement was suggested by Wissner-Gross which reviewed all the public advances over the last couple decades and found an evidence of what has been limiting the AI revolution for so long. His provocative explanation is that major breakthroughs have been limited by the presence and the quality of the relevant data, not by the small improvements in the algorithms itself.

TIMING OF BREAKTHROUGHS IN THE FIELD

YEAR	BREAKTHROUGHS	DATASET AVAILABLE	ALGORITHM PROPOSED
1994	Human-level spontaneous speech recognition	Spoken Wall Street Journal articles and other texts (1991)	Hidden Markov Model (1984)
1997	IBM DeepBlue defeated Garry Kasparov	700,000 Grandmaster chess games dataset (1991)	Negascout planning algorithm (1983)
2005	Google's Arabic- and Chinese-English translation	1.8 trillion tokens from Google Web and News pages (2005)	Statistical machine translation algorithm (1988)
2011	IBM Watson became the world Jeopardy! champion	8.6 million documents from Wikipedia and related projects (2010)	Mixture-of-Experts algorithm (1991)
2014	GoogLeNet object classification at near-human performance	ImageNet corpus of 1.5 million labeled images and 1000 objects (2010)	Convolutional Neural Network (1989)
2015	DeepMind achieved human parity in playing 29 Atari games by learning from video	Arcade Learning Environment dataset of over 50 Atari games (2013)	Q-learning algorithms (1992)

So, the average elapsed time between key algorithm proposals and corresponding advances was about 18 years, whereas the average elapsed time between key dataset availabilities and corresponding advances was less than three years, or about six times faster.

A perfect example of this theory is the Q-learning algorithm that was published more than twenty years ago but only had a chance to be used by Google DeepMind in 2015. Its modification achieved a human parity in playing twenty-nine Atari games was trained on the biggest gaming dataset called Arcade Learning Environment which consists of over 50 Atari games and was released in 2013.

The theory of dependency on high quality datasets may bring in order-of-magnitude improvements in machine intelligence breakthroughs. This means we may already have all the necessary algorithmic pieces to accomplish extremely complex tasks done by humans, though still lack a proper data for these algorithms to train on.

Having more data doesn't often mean bigger datasets, but better datasets with case-by-case exceptions when it comes down to the specific implementation and goals of the algorithms. Such an outlier case was described in Netflix paper which states that having just a few ratings was more helpful that having plenty of metadata.

In general situations, a mediocre deep learning algorithm trained on a huge amount of data will vastly outperform a great model trained on a small amount of data.

If this all sounds confusing, I recommend thinking of data as knowledge and an algorithm is used to digest this knowledge. If you have the best way to digest the knowledge but too little knowledge, you will still have very shallow understanding of the domain; if you have a lot of knowledge and a good way to digest it, you will still be able to know a lot.

Data also serves as a competitive advantage. Say you have a call center business - you have a much better chance of creating software that provides analysis of call center employee conversations. The other way would be having a strong partner or a client that is willing to share this data with you for the future benefits, but this is hard because of data privacy and security issues that partner may face.

Data is also a reason why it's a dead end to compete with huge giants in areas where they already have or can easily source related datasets. Most Google products are completely free because the entire business model is powered by the presence of data. Every Google product is designed to reduce friction in gaining data about you as a user, that's why you have free mail, free maps, free cloud storage and even a free operating system that is used by hundreds of smartphone manufacturers.

When it comes to sourcing the data, you can either use your own internal data if you have an existing dataset or source it externally.

So, where do you go when you are trying to source the data for your use case and don't have any of your own? The easiest answer is public datasets. They have been fueling most of the recent public innovation in the deep learning field. There are plenty of public datasets. One example is MNIST which is a dataset of 25x25 pixel, centered, B&W handwritten digits.

The more complex way to source external data is by scraping, and it may be quite expensive in terms of time spent on acquiring data and sometimes may not be fully legal.

If you can't find a source of public data for scraping, then you can try and synthesize the dataset. Online tools such as mechanical turk allow you to crowdsource the data by breaking it up into thousands of small micro jobs which may turn out to be quite expensive. This can range from fully generating the dataset or just labeling an existing dataset if the data is being used for supervised or semi-supervised learning.

There are also options of buying so-called commercial datasets. You can also license dataset from researchers or other companies doing research. Most likely you will be subject to the licensing terms by using it.

The other cool feature of deep learning architectures is the fact that you can easily build on top of other pre-trained networks using the called transfer learning method. Inspired by how humans transfer their learning, this method was created to leverage similarity behind most of the existing neural network structures.

A new somewhat shady way to obtain the dataset or existing pre-trained model is by reverse engineering the existing machine learning algorithm. Researchers have shown that given access to only a machine learning services it's possible to reverse-engineer machine learning algorithms with up to 99% accuracy. After discovering the actual algorithm, researchers were also able to generate examples of the potentially proprietary data from which it learned. Of course, the complexity of the algorithm directly represents how hard it is to reverse engineer. Simple algorithms that detect the presence of something with Yes or No can be deciphered quickly in contrast to more complex multidimensional algorithms.

Many companies that are trying to get on top of artificial intelligence often miss out the required level of strategy related to data. Others that understand the importance of latter get caught up in the hype of big data and start to collect as much information as possible, without paying attention to its relevancy to the real business needs.

The answer lies in the ability of businesses to leverage that data, apply new technologies and transform that into a competitive advantage. This leads to a strong argument that every business now is a data business and needs business strategy aligned with this need.

It all always starts with the business priorities and data strategy should always focus on those. It doesn't matter how often a customer calls customer service if your business question is how to target women who are expecting a baby. There is no point capturing terabytes of data that you don't really need, unless you are a huge giant such as Google or Amazon and you can afford it. Most companies never reach such scale thus lack resources and even organizational operations to fulfill those capabilities.

OVERCOMING FEARS AND MYTHS

It's a Hype

After seeing multiple periods of machine intelligence promises that came and failed, people are subject to a frequency illusion bias expecting to see the so-called period of AI winter happening again. The winter phenomenon is described as a cooling interest to the field after the rapid phase of progress and hype. The argument is obvious, if it happened so many times before, it will most probably happen again in some form or another.

But as the top researchers of the world keep saying, the improvements in computer processor design and microchip architectures will keep driving performance advances and breakthroughs for the decades ahead. Even though the current boom is much bigger than previous ones, the difference is in the expected commercial impact that is driving the trend.

In contrast to a dried-up government funding back in the 1970s, the current wave of researchers is well funded to perform fundamental research that already benefits big companies across the industry.

It's Too Early

Many people from the industry are claiming that it's too early to get the real benefits of AI or that we won't find anything useful until we hit the holy grail of so-called General Artificial Intelligence, the one that can do just about everything that humans do but better and faster, without specializing at one thing as many of the current implementations are.

The quest for AGI to be able to perform any intellectual task that a human being can is real and there is a lot of research going on there. It doesn't mean we must wait for that cross-domain expertise machines to start getting benefits in some niche areas.

It's Too Late

As much there are people saying that it's too early, there are also skeptics that think it's too late to try and build something valuable in the field because big companies are already doing exactly that. This might be true if you are aiming to create a horizontal type of machine intelligence that can serve the needs of the mainstream industry, such as creating a large-scale machine learning cloud like Amazon or Google.

There may be still potential if you are planning to target some very specific industry that is going to benefit from your machine learning cloud or if you tailor to the needs of specific customer segment that is underserved by big giants.

It's a Black Box

Even though most of the recent advances in deep learning are making it very hard to trace and explain the models that are producing the results, it still doesn't prevent businesses getting benefits from the use of machine learning. Of course, some areas may be trickier in use than others due to the specifics of results provided by AI.

There are arguments about the fact that no one can truly explain why certain algorithms and deep architectures work so good for the problems. Researchers have some ideas as to why a multi-layer neural network is more efficient than the ones with fewer layers, but overall the area is very experimental and is driven by the intuition and educated guesses. Most of the improvements in the field and industrial applications are results of intensive testing and experimentation. So, no one is stopping the work because they don't have enough theory as long as it provides real tangible benefits.

It's Not Applicable

Unless you have documented all your existing business processes, you can't really tell that it's not applicable to your company or business. The nature of businesses now is very wide, meaning that some areas of your organization can potentially produce never before seen processes if you start adopting machine intelligence across that areas. This fear can be easily debunked after a thorough examination of your existing business operations.

Lack of Experts and Resources

The fact that you are not an expert or don't have engineers in-house doesn't mean that you can't grasp the piece of the benefits provided by the AI revolution. This shouldn't be an excuse to not go out there and find expertise. Expertise doesn't even have to be in your location, or even in your own country. The benefits of remote distributed work are one of the underestimated pieces of the puzzle.

The lack of resources may be a very substantial risk if you are not doing things right. In some areas of machine intelligence application, you may require immense levels of funding and resources, but in most cases, you are only subject to properly identifying key areas and prioritizing the most needed development. The cost may be as low as paying a couple hundred dollars for a machine learning service that allows you to do your magic out of the box, if the business use case is not that complex.

CHAPTER FIVE

PREPARATION

BUSINESS PROCESSES

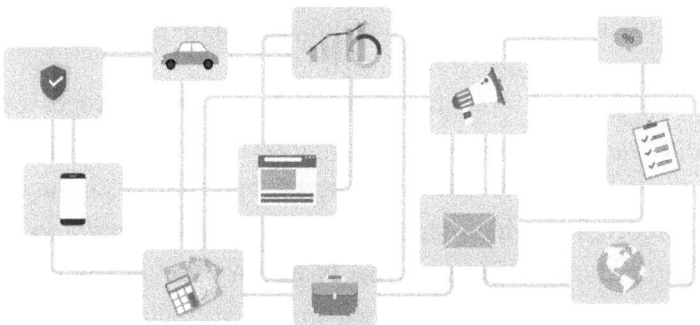

Once you understand the basic terminology behind AI press buzz and have digested the main principles of how things work and how they don't work, it's time to prepare for the actual use of these things.

There are two ways for a company to start using AI. The first one is when you are starting a company from scratch. Say you are an ambitious founder that decided to reinvent how a certain industry works but never had an actual business entity in the field.

The second one is much more interesting; it's the case when you already have a functioning business and want to transition into the use of AI. What I mean by more interesting is the fact that you already have established business processes that involve domain expertise and some form of data accumulation within the company, which you usually don't have when you are just starting out.

When you have an existing business and processes, you must write them out and formalize into easily digestible pieces or concepts. The purpose of this exercise is to map out your business in a way for the future analysis of where and how machine learning can help to improve it.

It's important to ensure that the documented processes work as described in your organization. In many cases, people tend to deviate from established routines in favor of shortcuts, which should be very well documented if that is the case because this is where some of the most beneficial insights may hide from the big picture.

When documenting the process, you need to clearly indicate what the overall goal of the process is and what are the metrics by which this goal is tracked. Also, you need to indicate what triggers the beginning of the process and what ends it.

Document all the activities at each step of the process including who is involved in them, whether it's a human, machine or a combination of both.

The nature of the business process itself assumes there is a certain interaction happening between the involved parties, which usually produces some outcomes. These outcomes are either in the form of data or can lead to the appearance of data somewhere down the funnel of the process.

You must clearly investigate and define areas that are able to accumulate data since these most likely are the easiest areas to automate depending on the nature of the interaction of course.

There are four types of interactions that produce data:

1. Human to Human (email, call, text)

2. Human to Machine (software operations)

3. Machine to Human

4. Machine to Machine (webhooks, notifications)

Even if the part of the business process currently doesn't produce any significant data, it may still transform into an area that does. These potential data collecting areas of the process are where the most valuable information is often generated.

When you start from scratch, you need to visualize the ideal business processes as you see it or as it works in similar organizations. This is much riskier since you are making a lot of assumptions on how things will work out and those assumptions is the core thing your strategy relies on.

After these processes are formalized and you have a clear picture of how your business works, you can start to formalize the data that is being accumulated.

IDENTIFY THE RIGHT AI STRATEGY

This section explains how to find the right strategy for your specific business.

After formalizing your processes, start thinking of the right strategy to apply AI. This obviously starts with the business use case and the end value proposition.

High-level business use cases can be divided into distinct techniques of:

- Optimization

- Identification

- Detecting anomalies

- Segmenting

- Enrichment

Optimization is the technique of reducing time or resources spent on a certain activity or process. Google maps traffic recognition feature is a perfect example how machine learning helps you get to a destination faster by taking into account potential delays caused by heavy traffic.

Identification is the technique of finding objects or specific data within the process area. Detecting all the clickbait articles that misuse human psychology to make people click on the articles within their social network feeds is a great example of how Facebook is trying to improve the quality of content within user feed.

Detecting anomalies is the technique of finding abnormal activity or outliers that don't conform to the expected behavior within a process or data generated by the process. Fraud detection, anti-hacking security and even tumor spotting on MRI scans are the examples of anomaly detection.

Segmenting is the technique of grouping the data or processes into meaningful and understandable segments for further use. Grouping users of a mobile app by their usage and their personal traits is the perfect example of segmenting.

Enrichment is the technique of making existing information more informable by adding or reformatting certain pieces of it. Augmentation of analytic data with the actionable information on what caused certain trends or deviations is the use of enrichment.

If you want to decrease time spent evaluating potential prospects for your company, use a combination of optimization, identification and enrichment. You can decrease time spent reviewing incoming leads by automatically identifying the low hanging fruit using machine learning and reduce time spent on doing research for that lead by enriching this lead with all the information available on the internet.

A side product of these techniques is the actual methods like classification, prediction and generation that help to achieve certain goals. Classification can serve as an example of identifying objects, faces in images or video, identifying letters, symbols or specific words. The prediction would give the ability to predict a probability of certain outcomes such as a customer will likely stop using your service or forecast demand for a product based on historical data.

Generation would give you the ability to fill in missing points in any type of multimedia or historical data.

Once you are familiar with high-level techniques, you must figure out what you want to achieve as a business. It doesn't matter what type of learning you select or what type of machine learning algorithm use if you are not sure what the business outcome would be.

The most successful companies usually solve a specific business problem and evolve over time into much wider solutions or platforms. The desire to experiment with technology makes it quite easy to get caught up in the algorithms and models and ignore the actual customer problems.

For most businesses, decision making is the easiest area to introduce machine learning. Decision making can be divided into two separate branches, automated decision making and decision support systems. Automated decision making eliminates the need for humans in the loop and most often only needs a formal acceptance for its decisions.

An example of this type of system is any marketing automation tool that sends out emails or messages based on specific criteria achieved by the user. For example, if a user added a product to a cart but hasn't checked out during the day, he should receive a follow-up email with recommended next steps.

If the process or objective is way too complicated with human-related nuances, and is not very clearly defined with the respect to the data, then the solution is to create decision support systems that can help teams become more effective than they are currently.

Some companies successfully use a combination of those two approaches. For example, DigitalGenius helps KLM Airlines handle the increasing volume of interactions with customers. Their intelligent application is trained on hundreds of thousands of previous customer-agent communications and proposes an auto-generated answer to a customer inquiry and the customer service agent decides whether to use it or not.

To begin improving any business decisions, you must start by formulating the right business questions since having clear objectives is extremely important to get the most benefits out of machine learning and automation.

These key unanswered questions will help you formulate and achieve your strategic goals. By focusing on these questions, it will be much easier to identify the core areas ripe for applying artificial intelligence.

The most important piece of your AI strategy should be the reason why you are striving to use AI. There are plenty of good reasons to start using machine learning, such as solving the real business problem by impacting the real world instead of academic research.

The most obvious bad reason is using technology for the sake of using it. Many businesses are trying to use the cool factor of AI and jump on the bandwagon of hype, but they will never succeed with a strategy of just doing AI and trying to find where to apply it. The screenshot of this job posting perfectly illustrates this issue. There is nothing worse than trying to apply technology to "something."

I own a successful e-commerce business and I want to use some of the profits from this business to fund and start an AI company. I also want to raise additional capital from VCs and angel investors.

I'm fascinated by AI and truly believe it is the future. I want to take existing AI technology, possibly modify it slightly, and apply it to something. I'm not even sure what yet. I was thinking maybe to help people find good deals in real estate. Maybe healthcare related. It could be anything. I'm open to suggestions.

I am looking for someone who is interested in taking a small salary (~$25,000) and equity (50%) to start this AI company with me. I'm 28 years old, a passionate entrepreneur, hard-working, and persistent. I'm a talented businessman and would handle everything on the business end. I need someone who is a talented AI developer.

Look forward to talking more,
John

Giants like Google can afford such experiments and can have such audacious goals. But you as a business should refrain from this strategy and avoid the goal of creating a so-called horizontal machine learning platform - the technology that is ubiquitous and can be applied to any industry. The goal for a successful machine learning company is to have a clear business problem to solve within a very niche vertical that no giant will ever touch. Luckily, there are plenty of such problems and opportunities in every single industry.

So, as any successful business, you should have a clear niche market to tackle that experiences a high value, frequently occurring problem. The more data rich this problem and the more clearly defined it is, the better. And of course, you need to have a product that is faster, better, or cheaper to carve out a decent portion of that market.

In general, the specifics of the AI industry are such that it can't be a cool side-project that you are willing to try out; it should have proper dedication in both resources and time. The fact that product market fit for AI companies is not any different from any other company just adds more dedication.

Of course, the goal of your AI strategy should be to focus on simple problems that are well defined and well understood, and where the available data exemplifies the information necessary to decide. This doesn't mean you need to use machine learning in areas where standard business logic can be applied; most benefits are usually hidden in areas where non-linear patterns or complex dependencies are presented.

If your customers are spread out across different industries, it's always better to focus on those where it will be the easiest to introduce your solution and get it into the hands of customers. Consider how much time and resources it will take your ideal customer to adopt the solution; sometimes the costs that are spent on a new solution may not have a quick payback period.

DATA PREPARATION

Once you are finished writing your business processes, you must formalize the data that is being generated within those processes. The first step is to indicate what type of data is generated or can be generated, how it is stored, and where it is stored.

If your business use case is centered around humans making the decision, then it is extremely important to collect as much data as is possible about the actual decision result, how that certain decision is made, along with all the scope of data that was used to make it.

The general process of preparing data can be broken down to three steps:

1. Extraction

2. Transformation

3. Loading

Extraction is the process of selecting the needed data from your various available sources. It can be stored in different databases, raw files on servers or even photos of documents or paper documents itself. At this point, you need to have a clear picture of what data is available and how to extract it. Excluding data is also important since having redundant data won't help you solve your problem and will complicate the further manipulations with it.

After extracting the data, you need to determine how this data will be used. Photos of documents or paper documents are not going to work, so you need to transform that data and establish the proper format that will be suitable for the future implementation. At this point, you may find out that some data is missing or is incomplete which should be addressed. There may also be sensitive data present in your data, which should be anonymized or removed if not necessary.

You may end up with more data than you need, which is always good, but you need to sample the appropriate amount of data that will be both enough to satisfy the needs and reduce the computational load needed to work with huge amounts of data.

The final step is loading, which means you must build the proper infrastructure for both storing that finalized data and making the retrieval as fast as possible. Data exploration and analysis can involve a lot of iterations, thus making the process of working with that data is very important.

As discussed in the previous chapter, you must have at least some data to begin. If you don't have any, and none of the options described previously are applicable, then you still have some options.

You can craft the dataset manually if you have the time and resources to do that. Of course, it depends on the type of the application you are exploring and the type of data that is needed. If no public dataset of traffic signs is available, you could easily go out yourself and spend a significant amount of time taking pictures of all the traffic signs in the city and labeling those. Or, you could hire people to do that for you. Either way, it will end up costing money and time, but if it makes economic sense, then why not?

The other strategy is to launch at least some form of the solution even if it doesn't include machine intelligence and begin collecting data. Depending on the nature of your application, you can either launch your early version with the human intelligence or no intelligence at all.

With human intelligence, you can serve a very small subset of users since you won't be able to scale that unless you are as big as Facebook which was exactly the case when they launched a beta version of Facebook M intelligent assistant that was powered manually by humans with the goal of teaching the algorithm.

When launching a solution with no intelligence at all, you must convince people to start using the application without getting machine learning benefits right away. That's how early adopters start using a product with the promise of intelligence to appear at some point. Usually this can be done by creating something that is already good at solving user's problem even without additional machine learning.

These products are data gathering by nature and collect that data through the user interaction with the system. Ideal examples are Google products that learn each time you use them, either through autocompleting searches, moving an email message to spam, or translating foreign sentences in Google translate.

The goal for any successful machine learning company is to build with the user in loop, meaning that the more users you have, the more data will be harnessed which will lead to smarter algorithms and eventually a better product. This phenomenon is called the *data network effect* and not only big giants are already exploiting the nature of it.

MORE USERS

BETTER PRODUCT

MORE DATA

IMPROVED MODEL

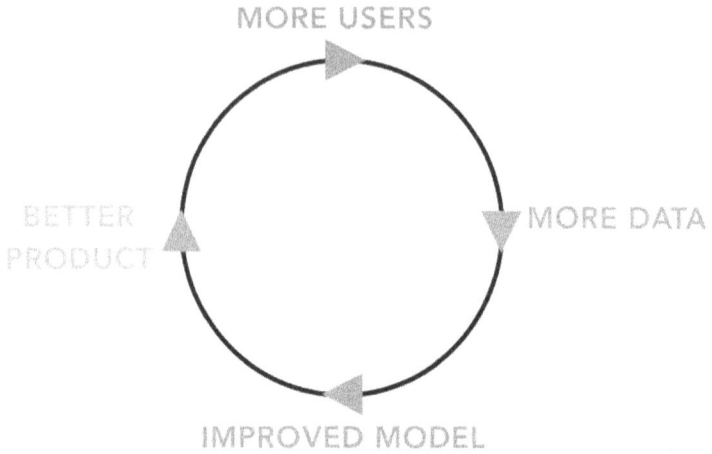

Similar to how Google products become better through time, Uber is becoming better at its routing algorithms due to an increase in transactions and that generates more data. It doesn't mean achieving data network effect is easy; it takes huge efforts, lots of data, and a proper infrastructure, but its potential is very exciting.

CHAPTER SIX

IMPLEMENTATION

This section explains how to implement the solution and start getting benefits

FINDING THE TALENT

No matter if you decide to use existing tools and machine learning services or decide to build your own, you will still need human talent, but there is a major difference in the type of talent you need to acquire. Usually, companies are stuck at this step which leads to very inefficient and wrong business decisions.

If you have no idea what technology to use or how to apply certain technology, how to source datasets and extract the needed expertise from your business, then you need a full stack company to help you. This can even be an agency; the key here is to pick a very specialized agency to what your needs are.

If your solution is more likely to be a computer vision application, then you need a CV-specialized agency or a proven team of people on the market. If your desired application domain is pure natural language analysis then, you need to find service companies that dream and live this industry and are able to showcase some very successful projects in the field.

Hiring a very generalized agency will lead to complete frustration and misuse of funds due to the specifics of this industry. Not that it's truly impossible to get results from a typical agency, but in most cases your chances on doing it efficiently and within the requested budget/timeline are extremely low.

You can try to construct such a specialized unit yourself, but you will need a lot of resources and expertise in many very narrow domains to do that. The first step would be to obtain a Chief Technology Officer (CTO), the person who is great at understanding technology at a high level and great at understanding the needs of the business.

There are tons of resources on how to hire great CTOs, but the specifics when dealing with machine learning are that this person must have at least some applicable knowledge in the domain, which many don't even if they've worked 5+ years as CTOs of Fortune 100 companies. The candidate doesn't necessarily need to code smart algorithms, but should have prior experience managing such teams and delivering the end result that brought benefits to the real business.

When diving deeper into technology talent, you should hire the actual executor who is a developer and is able to implement your requirements into the meaningful application. The talent crunch is real for the data science and machine learning industry and that will affect your hiring of the appropriate developer. Usually these visible and easy-to-reach top developers are concentrated in places like Silicon Valley though the competition there is fierce with the incredible salaries reaching the couple hundreds of thousands on average.

If your business has unlimited funding for the next AI project, then you can compete for that talent, but you must remember that those people are not money hungry and usually salaries work as the lowest denominator for the job. You need to provide the same or higher salary along with interesting or compelling projects.

The good thing is that it's easier than ever to hire talent globally. The rise of remote work and productivity tools enabling distributed work communication has created hundreds of companies where most employees work from different places in the world.

If you visit Kaggle rankings (the biggest platform that enables predictive and analytics competitions across the world), you'll see that the top machine learning engineers live all over the world, from Eastern Europe to Latin America. Kaggle was recently acquired by Google with the idea of democratizing AI research, but still most of the big giants hunt for machine learning talent over there and that is happening for a reason, there are over 600,000 engineers on the platform crunching through the toughest competitive problems in the world.

If your company has a fully-equipped technical department with its CTO and existing development teams, at some point you may need to introduce the missing piece of machine learning expertise. This means you are more likely to benefit from outsourcing that expertise with very detailed and clear requirements for what you need built to some company or a small team of specialized experts.

It may sound easier to get your existing team on track with machine learning and not to go somewhere else. Because of the current interest in the topic and availability of online education, many developers are more and more keen to learn the subject and tell you they've mastered machine learning at a desired level. In most cases, that level of knowledge is very sparse and can't compete with someone extremely tailored and specialized on the topic.

When this happens, companies spend more resources to build up expectations and get certain developers up to the needed level of expertise, which is very opposite to the real needs of the business - solving problems of customers. Of course, if you are ready to invest time and resources into creating T-shaped specialists this may be a good strategy, but most likely this is not true.

The modern availability of tools can create an illusion of this approach as a valid and it's very easy to fall into that trap. Keep that in mind the next time your web developer who just passed a Coursera course on machine learning shows you how his neural network (that was created with the open source tool) shows you some magic stuff like transforming his Instagram photos into Picasso-style drawings. Or your system administrator uses open source libraries or services to spot faces on the footage from your security camera.

These things seem to be a showcase of high-level knowledge, but they are not signs of any actual expertise in the domain of computer vision or machine learning. Spending a year trying to teach a general web developer how things work can cost your business a lot. The required task could take a month of work and cost less for a very specialized team that does this type of work on a daily basis. There are always edge cases and exceptions to these rules that are subject to your own risk.

One example of such an exception is when the problem of your business is very narrow, specific, and has been already solved by other companies in the field, moreover being available as a machine learning service on the Internet (for example, detecting NSFW content).

It would make no sense to create your own, the best in the world technology to detect adult content when there are existing services you can use for a very small cost compared to funding a whole R&D unit specialized on a very narrow topic.

An exception would be if you've reached a point in time where the scale of expenses associated with using the services is way higher than a year of salaries for an appropriate team to work on the problem. Most often that is not the case. Another exception might be if you are working in this industry and plan to compete with these services by creating your own solution that can be cheaper, faster and better at its job. It's hard to imagine that with the case of nipple detecting machine learning algorithms, but it may indeed make sense with some other very niche and industry-specific use cases.

For all other use cases, the number of machine learning services on the Internet is growing each day, including open source efforts and collaborative platforms where people are creating tools anyone can use for free.

BENEFITING FROM EXISTING SERVICES

There are plenty of services and existing infrastructure that may help your company identify the most fruitful way to benefit from AI. The question is if you want to achieve state-of-the-art efficiency and quality of the technique performing on your specific use case. If that 'snot the case, then you can benefit from existing platforms and infrastructure that offer these techniques in some form or another.

Even though machine learning services offered by other companies is not something new, many people in the field have been forecasting them to die off eventually. The hypothesis is that those types of services don't have a market. Since people that are specialized in machine learning and know what they are doing will just use open-source solutions and build something of their own. And people who don't know what they are doing won't be able to do anything even with APIs and ready-to-use services.

But that is only partially true. If you don't need a very fine-tuned machine learning solution, you can easily get by with just using existing service from companies who offer these machine learning services.

If your business application lies in the field of computer vision, you most likely will benefit from a range of online services offered by Clarifai company, which offers video and image recognition services. You can automatically tag all your images and videos for proper data organization that allows searching through your content. Moreover, visual recognition company Kairos is able to supercharge your application with things like face, age, gender and even emotion detection.

If your business needs to create a bot, most likely you don't need to reinvent the wheel and can benefit from Recast.ai collaborative platform that is used to build, train and deploy intelligent bots. You can leverage the existing infrastructure of the platform and connect your bots with hundreds of external services and monitor the performance of each.

If your business is audio related and needs to mine your data for some insights, keywords, or even simply transcribe your audio into text, then you can use DeepGram or VoiceBase suites of services exactly for that.

There are plenty of companies that can easily fulfill most of the basic needs when it comes to machine learning. Besides ready-to-use services, there are also open source tools and libraries that you as a company should be aware of. Libraries such as Tensorflow, Torch, Theano, Keras are all over the press for bringing the ease of deploying neural networks and democratizing the access to expertise when it comes to complex computation.

It's nearly impossible for a non-technical person to understand what these libraries are when you research them online. The explanation on the official site states: "TensorFlow is an open source software library for numerical computation using data flow graphs. Nodes in the graph represent mathematical operations, while the graph edges represent the multidimensional data arrays (tensors) communicated between them."

The only thing you need to understand is that it's much easier to build these complex architectures such as deep neural networks than it was five years ago and your engineers should take advantage of these libraries. If they don't, you are either aiming to create something better than the Googles of the world have achieved in this field, or simply wasting your time and resources. Treat those libraries as the pre-built building blocks for your machine learning algorithms.

COMMUNICATION WITH ENGINEERS

Anyone who has ever dealt with developers knows how hard it is not only to manage their work, but also to accurately communicate the business requirements, especially if the matter is as costly and error-prone as machine learning.

One common mistake senior management makes when dealing with machine learning developers is falling into the trap of avoiding communication. This tendency is forced by technical people, who in general try to avoid business people, meetings and discussions as much as possible. They like to code, optimize and geek out on their own and assume the communication is complete and won't happen until the fully developed solution is ready.

Do not make this mistake. Machine learning developers should understand that they are hired to solve business problems and without proper communication of these problems they won't be solved.

A good practice is a follow-up on the progress and scheduled reports on the high-level success of the project to collect feedback and potentially adjust the analysis and assumptions based on the feedback. This may sound very basic, but it is critical because wrong assumptions and premature optimizations can cost your business a fortune.

Another costly mistake is the ability of engineers to turn simple needs into super complex solutions. It may not always be obvious to the engineer that you as a business don't need to spend years on building the most complex architecture that will be drooled upon. The goal of any machine learning engineer is to solve customer's problem through an intelligent layer of the application, not build something that can handle tens of millions of users, unless explicitly requested to do so.

Sometimes this tendency to overcomplicate things also causes engineers to explore shiny new toys like deep learning when the use of much simpler algorithms will yield the same performance the complications involved with deep neural networks. Of course, you as a business owner should not understand these complex matters, but what you can do is to always force the decision and conversation towards what's the easiest and fastest way to solve this problem. Make sure you and your senior management are not the ones forcing the use of shiny new toys just because it's been everywhere in the press recently.

Delivering requirements is also an important topic that can cause tremendous amounts of time and resources being used incorrectly. Machine learning developers may be knowledgeable in the domain of neural networks and deep learning, but still fail to understand exactly how you want these things to be applied. This disconnect between business and AI talent is huge.

It is important that all the requirements are documented and managed by the responsible person, which usually is either someone delegated from senior management side or senior management itself if the company is quite small. One critical item that must be documented is the definition of success and how it looks like for a machine learning solution. The success is a quantitatively defined metric that shows a successful operation of the solution whether it is done by human or machine.

If there is no explicit requirement to what success is; it will be very hard for a machine learning engineer to construct any kind of system simply because most of them operate on the ability to match the intermediate results with the desired results and correct itself. Setting up evaluation metrics is a good way to tackle this problem. You must define which metrics are more important and which are less important.

The objectives are also important. Is the goal to do something better than human or at the same level as human? Or is the goal to assist a human operator at what he is doing by not being able to beat the performance? Knowing what the quantified performance of a human is helpful too in this case. These objectives along with metrics will influence the development of your solution.

DESIGNING FOR INTELLIGENCE

You can build the most advanced technology humanity has ever seen, but it is doomed if no one can figure out how to use it. The importance of good user interface design and user experience cannot be underestimated when dealing with AI.

The rule of thumb for a smart application is to make the use of machine intelligence almost invisible, like it doesn't even exist. The ordinary user of the system should not worry about which technology you are using or its limitations. A good design is the one that brings up the strong sides of your solutions and hides even obvious limitations.

The end result value proposition of your business process is much more important than the fancy interface that resembles robots and electrified brains that are the most popular AI visualizations.

Consider two examples of a smart application. The first one is filled with the buzzwords all across the user interface, has fancy robot characters all over the place, and emphasizes how smart it is and how much of the artificial intelligence it uses. The second one focuses on the value proposition, the problem that is being solved and the simplicity of the user interface versus fancy characters.

When it comes to the final step of the communication of the end user with the user interface, it is always the aha moment that makes your users love the product. That moment happens when the problem is solved in a timely manner and looks like magic, not when it is overhyped and not clear what happened, or didn't happen in the way as the end user expected. The process is similar to when a person tries Uber for the first time and magically gets a car in one minute without ever thinking about the computation and route optimization happening behind the scenes of the complex multivariable car matching mechanism.

That's why it's crucial that the designer understands both the goal of simplicity and delivering the unique value proposition and internal complexity of the system. This predicament is required to craft a fluid combination of both.

The complexity is often miscommunicated by developers and hurts the end user experience. Therefore, it's better to imagine that ideal magical moment first and iterate your way to the applicable viable version, rather than start crafting user experience based on the actual technical architecture and its limitations.

DEPLOYING SOLUTIONS

It is very important to set and communicate what the system you are deploying is capable of and what are the limitations. There is nothing worse than a vague value proposition filled with buzzwords that ruin the first-time user experience. The explicit communication and social engineering is a must when dealing with intelligent systems.

And of course, as much as the limits are important, you and your customers must acknowledge the right metrics and procedures of working with the system. You can't hope that a customer will come back after experiencing a cold start issue because their use hasn't generated enough value for the system to do the expected magic. If your solution is a subject to individual account cold start problem, when the user must use a system for some time to start generating insights, then you must explicitly indicate this and guide users through this situation and possible outcomes.

If your solution is a subject to gradual improvements in accuracy through time, you should manage the client's expectations and communicate that it will take some time for the results to improve.

Security and data privacy is another domain of concern when deploying intelligent data-fueled systems into the wild. The general objective of AI is extracting useful insights from data. Privacy should be preserved by concealing information so these two competing interests frequently balance when building machine intelligence. There is no single solution to this problem, though you as a business owner should always keep in mind the type of data you are dealing with and how sensitive it is to prevent future disputes.

It's also worth mentioning the topic of being stuck in never-ending development loop. This topic is specific to software companies and is one of the reasons why many companies fail without even getting their solution to the market. Much discussion around the agile methodology and lean startup perfectly applies to machine learning industry. It's very important to release your solution as fast as possible even if the solution is not perfect. Of course, it shouldn't hurt the quality and inability to fulfill the value proposition but in general it's better to release at least something to see the feedback coming in sooner.

CONCLUSION

No one can predict how will the world look like a decade from now, but one can tell is that it will be a completely different world. AI is a foundational technology that will power that change. It is important to emphasize that education must occur across industries to adopt this technology in ways that will be beneficial to society.

For many organizations, the required level of change invokes fear and pessimism. It's not only technology that has to be adopted, but a whole new way of thinking through the prism of machine intelligence. From properly inspecting your customer needs to collecting relevant data, the theory behind the successful implementation is still very new and subject to many changes, as the fast pacing nature of this technology is.

Through helping dozens of early stage startups and many more traditional companies unlock the power of AI, I've learned that following a structural framework is key to implementation when dealing with machine intelligence. This exact framework is laid out as a book and designed to be easily read. The purpose of this book is not a highly-detailed exploration of technology, but rather a simplistic view to help readers understand the landscape and navigate it more efficiently. And I truly hope the goal is accomplished successfully.

To follow me and my journey of helping companies, join the invite-only community of people interested in AI and its implications by going to www.algorology.com and entering your email. You will receive a weekly newsletter on what's new and interesting in the world of machine intelligence and will be personally welcomed by me, not subject to machine automation when it comes to personal relationships. I will also send you a list of useful AI tools you can start using in your life.

Another way of connecting is to visit my personal site at ArturKiulian.com, add me on Facebook or follow my Twitter.

ABOUT THE AUTHOR

Artur Kiulian is a serial entrepreneur, speaker, and author. He holds Master's degree in Systems of Artificial Intelligence and is a well-published writer on topics of AI and Machine Learning. As a Partner at Los Angeles based venture studio Colab, he is helping early stage startups figure out what software products to build and how to get them to the market as fast as possible.

ADDITIONAL RESOURCES

MACHINE-LEARNING-AS-A-SERVICE PLATFORMS

GENERAL

IBM Watson
https://developer.ibm.com/watson/

Google Cloud Machine Learning Engine
https://cloud.google.com/products/machine-learning/

Amazon Machine Learning
https://aws.amazon.com/machine-learning/

Microsoft Azure Machine Learning
https://azure.microsoft.com/en-us/services/machine-learning/

BigML
https://bigml.com/

H2O
https://www.h2o.ai

Prediction.io
https://www.prediction.io

CrowdFlower
https://www.crowdflower.com/

DataRobot
https://www.datarobot.com/cloud/

COMPUTER VISION

Amazon Rekognition
https://aws.amazon.com/rekognition/

Google Cloud Vision
https://cloud.google.com/vision/

Cloudsight
https://cloudsight.ai/

Clarifai
https://www.clarifai.com/

Kairos
http://kairos.com/

IBM Watson Visual Recognition API
https://www.ibm.com/watson/developercloud/visual-recognition/

Cognitec
http://www.cognitec.com/

AUDIO

Google Cloud Speech
https://cloud.google.com/speech/

Twilio Speech Recognition
https://www.twilio.com/speech-recognition

Nexmo Voice API
https://www.nexmo.com/products/voice

Microsoft Azure Cognitive Services
https://azure.microsoft.com/en-us/services/cognitive-services

VoiceBase
https://www.voicebase.com/

DeepGram
https://www.deepgram.com/

LANGUAGE

Rosette
https://www.rosette.com/

IBM Watson
https://www.ibm.com/watson/developercloud/nl-classifier.html

TextRazor
https://www.textrazor.com/

Google Cloud Natural Language
https://cloud.google.com/natural-language/

Lexalytics
https://www.lexalytics.com/

Indico.io
https://indico.io

CONVERSATIONAL BOTS

Amazon Lex
https://aws.amazon.com/lex/

IBM Watson Conversation Service
https://www.ibm.com/watson/developercloud/conversation.html

AgentBot
http://agentbot.net/en/

Twyla
https://www.twylahelps.com/

Msg.ai
http://msg.ai/

wit.ai
https://wit.ai/

Api.ai
https://api.ai/

Microsoft Bot Framework
https://dev.botframework.com/

Chatfuel
https://chatfuel.com/

Octane.ai
https://octaneai.com/

ManyChat
https://manychat.com/

Reply.ai
https://www.reply.ai/

Datasets
https://en.wikipedia.org/wiki/List_of_datasets_for_ma
chine_learning_research

https://aws.amazon.com/public-datasets/

https://github.com/caesar0301/awesome-public-
datasets

OBJECT RECOGNITION

**DAVIS (Densely Annotated VIdeo
Segmentation): dataset contains videos of
several types of objects and humans with a high-
quality segmentation annotation**
http://davischallenge.org/

T-LESS: dataset and evaluation methodology for detection and 6D pose estimation of texture-less objects
http://cmp.felk.cvut.cz/t-less/

Berkeley 3-D Object Dataset contains 849 images taken in 75 different scenes. About 50 different object classes are labeled.
http://kinectdata.com/

Berkeley Segmentation Data Set and Benchmarks 500 (BSDS500): an extended version of the BSDS300 that includes 200 fresh test images
https://www2.eecs.berkeley.edu/Research/Projects/CS/vision/bsds/

Microsoft Common Objects in Context (COCO): database of new image recognition, segmentation, and captioning dataset.
http://mscoco.org/

SUN Database: comprehensive collection of annotated images covering a large variety of environmental scenes, places and the objects within
http://groups.csail.mit.edu/vision/SUN/

ImageNet: labeled object image database
http://www.image-net.org/

TV News Channel Commercial Detection Dataset
http://archive.ics.uci.edu/ml/datasets/tv+news+channe
l+commercial+detection+dataset

Statlog: an image segmentation database
https://archive.ics.uci.edu/ml/datasets/Statlog+(Image
+Segmentation)

Caltech 101: pictures of objects belonging to 101 categories
http://www.vision.caltech.edu/Image_Datasets/Caltech
101/

Caltech-256: large dataset of images for object classification
http://authors.library.caltech.edu/7694/

LabelMe: annotated pictures of scenes
http://labelme.csail.mit.edu/Release3.0/browserTools/
php/dataset.php

Cityscapes Dataset: contains a diverse set of stereo video sequences recorded in street scenes from 50 different cities
https://www.cityscapes-dataset.com/

PASCAL VOC Dataset: database contains large number of images for classification tasks.
http://host.robots.ox.ac.uk/pascal/VOC/

CIFAR-1: Dataset consists of 60000 32x32 colour images in 10 classes, with 6000 images per class
https://www.cs.toronto.edu/~kriz/cifar.html

CIFAR-100 Dataset: is just like the CIFAR-10, except it has 100 classes containing 600 images each.
https://www.cs.toronto.edu/~kriz/cifar.html

German Traffic Sign Detection Benchmark Dataset: data set of over 50,000 images of 43 different street signs
http://benchmark.ini.rub.de/

KITTI Vision Benchmark Dataset: contains images of various areas using cameras and laser scanners.
http://www.cvlibs.net/datasets/kitti/eval_odometry.php

HANDWRITING AND CHARACTER RECOGNITION

MNIST: dataset of 25x25, centered, B&W handwritten digits.
https://pjreddie.com/projects/mnist-in-csv/

CIFAR: 32x32 color images.
https://www.cs.toronto.edu/~kriz/cifar.html

Artificial Characters Database: artificially generated data describing the structure of 10 capital English letters
https://archive.ics.uci.edu/ml/datasets/Artificial+Characters

Letter Dataset: upper case printed letters
http://archive.ics.uci.edu/ml/support/letter+recognition

Character Trajectories Dataset: multiple, labelled samples of pen tip trajectories recorded whilst writing individual characters
https://archive.ics.uci.edu/ml/datasets/Character+Trajectories

UJI Pen Characters Dataset: more than 11k isolated handwritten characters
https://archive.ics.uci.edu/ml/datasets/UJI+Pen+Characters+(Version+2)

Gisette Dataset: handwriting samples from the often-confused 4 and 9 characters
https://archive.ics.uci.edu/ml/datasets/Gisette

MNIST Database: database of handwritten digits
http://yann.lecun.com/exdb/mnist/

Optical Recognition of Handwritten Digits Dataset: database of known letters/digits
http://archive.ics.uci.edu/ml/datasets/optical+recognition+of+handwritten+digits

Pen-Based Recognition of Handwritten Digits Dataset: database of 250 handwritten digit samples from 44 writers
https://archive.ics.uci.edu/ml/datasets/Pen-Based+Recognition+of+Handwritten+Digits

Semeion Handwritten Digit Dataset: handwritten digit dataset from 80 people
http://archive.ics.uci.edu/ml/datasets/semeion+handwritten+digit

HASYv2: database of handwritten mathematical symbols
https://www.semanticscholar.org/paper/The-HASYv2-dataset-Thoma/f6c0f6722f78b68ad99a0b648a6b61b0fb524a06

ACTION RECOGNITION

Human Motion DataBase (HMDB51): Large Video Database for Human Motion Recognition
https://link.springer.com/chapter/10.1007/978-3-642-33374-3_41

TV Human Interaction Dataset: contains videos from 20 different TV shows for prediction social actions: handshake, high five, hug, kiss and none
http://www.robots.ox.ac.uk/~alonso/tv_human_interactions.html

UT Interaction: dataset contains videos of continuous executions of 6 classes of human-human interactions: shake-hands, point, hug, push, kick and punch
http://cvrc.ece.utexas.edu/SDHA2010/Human_Interac tion.html

UT Kinect: dataset was collected as part of research work on action recognition from depth sequences
http://cvrc.ece.utexas.edu/KinectDatasets/HOJ3D.html

SBU Interact: database contains information about seven participants performing one of 8 actions together (approaching, departing, pushing, kicking, punching, exchanging objects, hugging, and shaking hands) in an office setting
http://www3.cs.stonybrook.edu/~kyun/research/kinect _interaction/index.html

Berkeley Multimodal Human Action Database (MHAD): contains 11 actions performed by 7 male and 5 female subjects in the range 23-30 years of age except for one elderly subject.
http://tele-immersion.citris-uc.org/berkeley_mhad

UCF 101 Dataset (Action Recognition Data Set): data set of realistic action videos, collected from YouTube, having 101 action categories.
http://crcv.ucf.edu/data/UCF101.php

THUMOS Dataset: large video dataset for action classification.

http://www.thumos.info/home.html

Activitynet: large video dataset for activity recognition and detection

http://activity-net.org/

MSP-AVATAR: is a motion capture database which explores the role of discourse functions in non-verbal human interactions

http://ecs.utdallas.edu/research/researchlabs/msp-lab/MSP-AVATAR.html

AERIAL IMAGERY

Aerial Image Segmentation Dataset: 80 high-resolution aerial images with spatial resolution ranging from 0.3 to 1.0.

https://project.inria.fr/aerialimagelabeling/

KIT AIS Data Set: comprises aerial image sequences and a xml files with manually labeled trajectories of all visible vehicles

https://www.ipf.kit.edu/english/downloads_707.php

Wilt Dataset: consists of remote sensing data of diseased trees and other land cover

https://archive.ics.uci.edu/ml/datasets/Wilt

Forest Type Mapping Dataset: contains satellite imagery of forests in Japan
https://archive.ics.uci.edu/ml/datasets/Forest+type+mapping

Overhead Imagery Research Data Set: consist of annotated overhead imagery
https://sourceforge.net/projects/oirds/

REVIEWS

Amazon reviews: database of US product reviews from Amazon.com.
http://jmcauley.ucsd.edu/data/amazon/

OpinRank Review Dataset: contains full user reviews for cars collected from Edmunds.com. Hotel reviews from TripAdvisor.com are also included but without user ID's in the same link.
http://kavita-ganesan.com/entity-ranking-data

MovieLens: database has 22,000,000 ratings and 580,000 tags applied to 33,000 movies by 240,000 users
https://movielens.org/

Yahoo! Music User Ratings of Musical Artists: database contains over 10M ratings of artists by Yahoo users
https://webscope.sandbox.yahoo.com/catalog.php?datatype=r

Car Evaluation Data Set: contains car properties and their overall acceptability
https://data.world/uci/car-evaluation

YouTube Comedy Slam Preference Dataset: contains user vote data for pairs of videos shown on YouTube
https://archive.ics.uci.edu/ml/datasets/YouTube+Comedy+Slam+Preference+Data

Skytrax User Reviews Dataset: contains user reviews of airlines, airports, seats, and lounges from Skytrax
http://www.airlinequality.com/

Teaching Assistant Evaluation Dataset: contains teaching assistant reviews
https://archive.ics.uci.edu/ml/datasets/teaching+assistant+evaluation

NEWS

The Reuters Corpus Volume 1: database of large corpus of Reuters news stories in English
http://www.daviddlewis.com/resources/testcollections/rcv1/

The Reuters Corpus Volume 2: database of large corpus of Reuters news stories in multiple languages.
https://archive.ics.uci.edu/ml/datasets/Reuters+RCV1+RCV2+Multilingual,+Multiview+Text+Categorization+Test+collection

Thomson Reuters Text Research Collection: large corpus of news stories.
http://livingarchive.inn.ac/datasets/show/51125eeac7c2f7779a500a00

Saudi Newspapers Corpus(SaudiNewsNet): contains a set of 31,030 Arabic newspaper articles along with metadata, extracted from various online Saudi newspapers
https://github.com/ParallelMazen/SaudiNewsNet

CORRESPONDENCE

Enron Email Dataset: contains emails from employees at Enron organized into folders
https://www.cs.cmu.edu/~./enron/

Ling-Spam Dataset: corpus containing both legitimate and spam emails
http://csmining.org/index.php/ling-spam-datasets.html

SMS Spam Collection Dataset: collection of SMS spam messages
https://www.kaggle.com/uciml/sms-spam-collection-dataset

Twenty Newsgroups Dataset: contains messages from 20 different newsgroups
http://qwone.com/~jason/20Newsgroups/

Spambase Dataset: contains spam emails
https://archive.ics.uci.edu/ml/datasets/spambase

Sentiment140: dataset contains tweets from 2009 including original text, time stamp, user and sentiment
http://www.sentiment140.com/

ASU Twitter Dataset: Twitter network data, not actual tweets. Shows connections between a large number of users
http://socialcomputing.asu.edu/datasets/Twitter

SNAP Social Circles: large twitter network data.
https://snap.stanford.edu/data/

Buzz in Social Media Dataset: consist of data from Twitter and Tom's Hardware. This dataset focuses on specific buzz topics being discussed on those sites
http://ama.liglab.fr/resourcestools/datasets/buzz-prediction-in-social-media/

Paraphrase and Semantic Similarity in Twitter (PIT): dataset is focuses on whether tweets have (almost) same meaning/information or not
https://www.researchgate.net/publication/301446243_
Ebiquity_Paraphrase_and_Semantic_Similarity_in_Tw
itter_using_Skipgrams

AUDIO

Zero Resource Speech Challenge 2015: database consists of spontaneous speech (English), Read speech (Xitsonga).
http://sapience.dec.ens.fr/bootphon/

Parkinson Speech Dataset: multiple recordings of people with and without Parkinson's Disease
https://archive.ics.uci.edu/ml/datasets/Parkinson+Spe
ech+Dataset+with++Multiple+Types+of+Sound+Recor
dings

ISOLET Dataset: dataset of spoken letter names
https://archive.ics.uci.edu/ml/datasets/ISOLET

Parkinson's Telemonitoring Dataset: multiple recordings of people with and without Parkinson's Disease
https://archive.ics.uci.edu/ml/datasets/parkinsons+tele
monitoring

TIMIT: recordings of 630 speakers of eight major dialects of American English, each reading ten phonetically rich sentences.
https://catalog.ldc.upenn.edu/ldc93s1

Geographical Original of Music Data Set: audio features of music samples from different locations
http://archive.ics.uci.edu/ml/datasets/geographical+original+of+music

Million Song Dataset: is a freely-available collection of audio features and metadata for a million contemporary popular music tracks
https://labrosa.ee.columbia.edu/millionsong/

Free Music Archive: database consist of audio under Creative Commons from 100k songs (343 days, 1TiB) with a hierarchy of 161 genres, metadata, user data, free-form text.
https://github.com/mdeff/fma

Bach Choral Harmony Dataset: bach chorale chords.
https://archive.ics.uci.edu/ml/datasets/Bach+Choral+Harmony

UrbanSound: labeled sound recordings of sounds like air conditioners, car horns and children playing
https://serv.cusp.nyu.edu/projects/urbansounddataset/

SENSORS

Cuff-Less Blood Pressure Estimation Dataset: cleaned vital signals from human patients which can be used to estimate blood pressure
https://archive.ics.uci.edu/ml/datasets/Cuff-Less+Blood+Pressure+Estimation

Gas Sensor Array Drift Dataset: measurements from 16 chemical sensors utilized in simulations for drift compensation
https://archive.ics.uci.edu/ml/datasets/Gas+Sensor+Array+Drift+Dataset+at+Different+Concentrations

Servo Data set: dataset covering the nonlinear relationships observed in a servo-amplifier circuit
http://archive.ics.uci.edu/ml/datasets/servo

UJIIndoorLoc-Mag Dataset : indoor localization database to test indoor positioning systems
http://archive.ics.uci.edu/ml/datasets/UJIIndoorLoc-mag

Sensorless Drive Diagnosis Dataset: electrical signals from motors with defective components
https://archive.ics.uci.edu/ml/datasets/dataset+for+sensorless+drive+diagnosis

Wearable Computing: Classification of Body Postures and Movements (PUC-Rio): people performing five standard actions while wearing motion tackers.
https://archive.ics.uci.edu/ml/datasets/Wearable+Computing%3A+Classification+of+Body+Postures+and+Movements+(PUC-Rio)

Gesture Phase Segmentation Dataset: features extracted from video of people doing various gestures
https://archive.ics.uci.edu/ml/datasets/gesture+phase+segmentation

Vicon Physical Action Data Set Dataset: 10 normal and 10 aggressive physical actions that measure the human activity tracked by a 3D tracker
https://archive.ics.uci.edu/ml/datasets/Vicon+Physical+Action+Data+Set

Human Activity Recognition Using Smartphones Dataset: gyroscope and accelerometer data from people wearing smartphones and performing normal actions.
https://archive.ics.uci.edu/ml/datasets/human+activity+recognition+using+smartphones

Australian Sign Language Signs: motion-tracking gloves data.
https://archive.ics.uci.edu/ml/datasets/Australian+Sign+Language+signs

Weight Lifting Exercises monitored with Inertial Measurement Units: dataset of five variations of the biceps curl exercise monitored with IMUs.
https://archive.ics.uci.edu/ml/datasets/Weight+Lifting +Exercises+monitored+with+Inertial+Measurement+U nits

sEMG for Basic Hand movements Dataset: two databases of surface electromyographic signals of 6 hand movements
https://archive.ics.uci.edu/ml/datasets/sEMG+for+Basi c+Hand+movements

MACHINE LEARNING TOOLKITS

Tensorflow
https://www.tensorflow.org/

Keras
https://keras.io/

Caffe
http://caffe.berkeleyvision.org/

Torch
http://torch.ch/

Apache Mahout
http://mahout.apache.org/

Apache Singa
https://singa.incubator.apache.org/en/index.html

Theano
http://deeplearning.net/software/theano/

Veles
https://velesnet.ml/

Scikit-Learn
http://scikit-learn.org/

Deeplearning4j
https://deeplearning4j.org/

Lasagne
https://lasagne.readthedocs.io/en/latest/

CNTK
https://github.com/Microsoft/CNTK

ONLINE COURSES

UC Berkeley: Intro to Artificial Intelligence
http://ai.berkeley.edu/home.html

Machine Learning (Stanford University via Coursera)
https://www.coursera.org/learn/machine-learning

Learning from Data - Abu Mostafa, EdX and Caltech
https://www.edx.org/course/learning-data-introductory-machine-caltechx-cs1156x-0

Udacity: Machine Learning Intro
https://www.udacity.com/course/intro-to-machine-learning--ud120

Udacity: Machine Learning Free Course
https://www.udacity.com/course/machine-learning--ud262

Udacity: Machine Learning Engineer Nanodegree
https://www.udacity.com/course/machine-learning-engineer-nanodegree--nd009

Machine Learning (Columbia University via edX)
https://www.edx.org/course/machine-learning-columbiax-csmm-102x-1

Machine Learning A-Z™: Hands-On Python & R In Data Science (Kirill Eremenko, Hadelin de Ponteves, and the SuperDataScience Team via Udemy)
https://www.udemy.com/machinelearning/

Machine Learning Specialization: Build Intelligent Applications
https://www.coursera.org/specializations/machine-learning

Principles of Machine Learning
https://www.edx.org/course/principles-machine-learning-microsoft-dat203-2x-4

Machine Learning for Data Analysis
https://www.coursera.org/learn/machine-learning-data-analysis

Practical Predictive Analytics: Models and Methods
https://www.coursera.org/learn/predictive-analytics

Master Recommender Systems Specialization
https://www.coursera.org/specializations/recommender-systems

Machine Learning for Musicians and Artists
https://www.kadenze.com/courses/machine-learning-for-musicians-and-artists/info

Machine Learning for Data Science
https://www.edx.org/course/machine-learning-data-science-uc-san-diegox-ds220x

Neural Networks for Machine Learning
https://www.coursera.org/learn/neural-networks